Fruit Production

Fruit Production
Theory and Practicals

Theory Part by
K. Usha
Madhubala Thakre
Amit Kumar Goswami
Nayan Deepak, G.
Practicals Part by
Anil Kumar Dubey
Om Praksh Awasthi
Sanjay Kumar Singh
V.B. Patel

Division of Fruits and Horticultural Technology
ICAR- Indian Agricultural Research Institute
(ICAR-IARI) Pusa, New Delhi – 110 012

NEW INDIA PUBLISHING AGENCY
New Delhi – 110 034

NEW INDIA PUBLISHING AGENCY
101, Vikas Surya Plaza, CU Block, LSC Market
Pitam Pura, New Delhi 110 034, India
Phone: + 91 (11)27 34 17 17 Fax: + 91(11) 27 34 16 16
Email: info@nipabooks.com
Web: www.nipabooks.com

Feedback at feedbacks@nipabooks.com

© 2020, Authors

ISBN : 978-93-89130-18-8

Composed and Designed by NIPA

Preface

India has diverse climates and soil for cultivation of horticulture crops providing ample opportunities for the development of fruit industry but the greatest challenge in present is to produce sufficient fruit to feed the ever-increasing human population. This demands infusion of technology for an efficient utilization of resources for deriving higher output per unit of input with excellent quality of the produce within short span of time.In the present era of open economy; it has become increasingly necessary that our produce is competitive, both in the domestic as well as in international markets. This would be possible only through deployment of high-tech application and precision farming methods.

Fruits are very important for human beings and also play an important role in religious practices, mythology and art. They are not only delicious but also have many nutrients which are necessary for human health. India is the second largest producer of fruits in the world. Fruit production requires a lot of science and some basic fundamental knowledge to grow them successfully. This book compiles many fundamental issues of fruit production like layout and planting, many cultural practices, growth and bearing habits of fruit crops, unfruitfulness etc. Understanding of all these topics will help the students for proper knowledge of fundamentals of fruit production.

We hope that this book would also be helpful for growers, nurserymen, farmers, teachers, scientist, extension officer and all those who wish to become familiar with the topic in relation to their professional interest. We have tried to keep the language as simple and straightforward as possible and consistent with accurate representation of the content. Every effort has been made to present the ideas in very easy and understandable language and the interests of each reader. Finally, we would like to thank one and all who have contributed directly or indirectly in bringing out this book.

Authors

Contents

Part-I: Theory
K. Usha, Madhubala Thakre, Amit Kumar Goswami and Nayan Deepak, G.

Part-II: Practicals
Anil Kumar Dubey, Om Praksh Awasthi, Sanjay Kumar Singh and V.B. Patel

Colour Section of Part-I: Theory

Colour Section of Part-II: Practicals

Part-I: Theory

K. Usha, Madhubala Thakre, Amit Kumar Goswami and Nayan Deepak, G.

1

Orchard Layout and Establishment of Fruit Orchard – Principles and Practices

K. Usha

Division of Fruits and Horticultural Technology, ICAR-IARI, New Delhi, India

Growing fruits can be fun, delightful experience and a challenging hobby. There are several good reasons to grow fruits. Fruits add beautiful colour and variety to your garden. Large amounts of fruit can be grown in a relatively small area.Store-bought fruits are often picked, shipped, and sold before they are fully ripened and generally stock selections that look prettiest but are not necessarily the best tasting. In addition, fresh garden fruits are a good source of vitamins; minerals, carbohydrates, fibre (Table 1) and will provide your family with flavorful, delicious and nutritious food. As an added bonus, the fruits you grow will taste much better than the fruits you find in the grocery store. Establishment of an orchard are a long term investment and require critical planning, selection of proper location and site, planting system and planting distance, varieties and the nursery plants to ensure maximum production.

Even if you have limited space, you can still enjoy growing fresh fruits in your garden by growing fruit plants in containers. You can grow any fruit tree in a container for a few years and then transplant it. You can also choose a dwarf variety, which is well suited to living in a container. If you carefully choose the kind and varieties of a fruit before you plant, you can harvest dessert-quality fruit from early summer through the fall. Many fruit plants are aesthetically pleasing and are good for "edible landscaping." Some fruits, such as apple, citrus, blueberries and figs, make outstanding ornamental plants. Some people find a natural setting with plants placed throughout the landscape most aesthetically pleasing. But this makes maintenance more challenging.

How to choose plants for your orchard?

There are so many different kinds of fruits and varieties available. How then to decide which to grow? The type and mix of species chosen depends on food preferences, soil and climatic conditions, availability of local materials and seeds. Start with quality. When soft berries are home-grown, flavour is outstanding and they can be harvested when fully ripe, plump, and sweet, without concern for shipping and ability to perish. The amount of yard space available will be another deciding factor. Most fruit trees come in varieties that are dwarf, semi-dwarf, and standard. Dwarf trees stay about 8- to 10-feet tall and are perfect for small yards. There are even dwarf fruit trees, such as columnar apples, figs, and pomegranates that fit well in containers, allowing you to move them as needed. Semi-dwarf trees grow about 10- to 15-feet tall and are more productive than dwarfs. Standard-sized trees grow 20 feet or more and are most productive. Sketch out how much room you have in your garden to determine which type of fruit tree you should plant. Another consideration is the size of the tree. Choose between growing small fruits - berries that grow on small plants, vines, or bushes - or larger tree fruits. Small fruit trees especially dwarf and semi dwarf varieties (Table 3) are the most exotic plants you can choose for home garden as they require minimum space and come to bearing within one or two years after planting. Their lush foliage, exotic flowers and edible fruits make captivating and enchanting home garden. Also, it is convenient to do cultural and pest control operations. Depending on water availability, gardening may be practiced year-round or limited to seasons. The length of your growing season is also important, as some cultivars require a long frost-free period to ripen fruit. Some also require a minimum number of "growing degree days" (GDDs), essentially a measure of how much warm weather you have at your site. Choose fruit plants accordingly.

Now narrow down your choices to fit the space available. The first rule is, plant only fruits that you would like to eat. Then select fruits that are most suited to your area, climate (Table 2), as this will give you minimal disease problems and a better yield. Another consideration is the variety of fruit tree you're growing. When choosing cultivars, look for those with outstanding hardiness, disease resistance, and fruit quality. Many of the newer cultivars provide top quality fruits not often available in local markets. You should not plant strawberries or raspberries where crops that are susceptible to *Verticillium* wilt have been grown (these include potatoes, tomatoes, eggplants, and peppers). Choose cultivars resistant to *Verticillium* wilt. Some fruits are easier to grow whilesome fruit plants require a lot of care and don t come into full production for several years. Strawberries aren t much harder to grow and bear fruit quickly. Most of the other fruit trees such as mango, citrus, grape, papaya, banana, apples, peaches, and pears require some knowledge and attention to pollination, pruning, spraying, fertilizing, and other kinds of care. While consulting a professional is often the best course of action to resolve any serious fruit tree problems, knowing few basics about fruit tree care can save you money and give you the satisfaction of caring your own fruit garden. Hence, carefully consider selection of fruit plants, design of your planting, including arrangement,

Table 1: Approximate nutrient composition (per 100 g) of various fruits

Fruit crop	Calories (kcal)	Carotene Vit.A (IU)	Vit.B1 (mg)	Vit.B2 (mg)	Niacin (mg)	Vit.C (mg)	Ca (mg)	P(mg)	Fe(mg)	K(mg)
Almond	655	trace	240	-	2.5	-	0.23	0.49	3.5	-
Aonla	59	-	30	-	0.2	600	0.05	0.02	1.2	-
Apple	56	Trace	120	30	0.2	2	0.01	0.02	1.7	144
Apricot	201	98	217	-	2.2	trace	-	-	0.04	-
Avocado	215	-	-	-	-	13	0.01	0.08	0.7	-
Bael	129	186	12	1191	0.9	15	0.09	0.05	0.3	-
Banana	153	trace	150	30	0.3	1	0.01	0.05	0.4	396
Ber	55	70	-	-	-	160.0	0.03	0.03	0.8	-
Blueberry	62.4	-	-	-	-	22.0	10.0	9.1	0.74	65
Breadfruit	79	15	-	-	-	-	0.04	0.03	0.5	-
Capegooseberry	55	-	-	-	-	49	0.01	0.06	1.8	-
Cherry	60.2	-	-	-	-	12.0	8.0	7.0	-	114
Currant	45.0	-	-	-	-	36.0	29.0	27.0	0.91	238
Custard apple	105	trace	-	-	-	-	0.02	0.04	1.0	-
Date	283	600	90	30	0.8	trace	0.07	0.08	10.6	-
Elderberry	46.4	-	-	-	-	18.0	35.0	57.0	-	305
Fig	75	270	-	50	0.6	2	0.06	0.03	1.2	194
Grape	45	15	40	10	0.3	3	0.03	0.02	0.4	0.15
Grape fruit	45	-	120	20	0.3	31	0.03	0.03	0.2	-
Guava	66	trace	30	30	0.2	299	0.01	0.04	1.0	-
Jackfruit	84	540	30	30	0.4	-	0.02	0.03	0.5	-
Jamun	83	-	-	-	-	10	0.02	0.01	1.0	-
Karonda	364	-	-	-	-	-	0.16	0.06	39.1	-
Lemon	57	trace	89	4	0.1	60.0	0.07	0.01	2.3	-
Lime	59	26	20	-	0.1	63.0	0.09	0.02	0.3	-

Contd.

Fruit crop	Calories (kcal)	Carotene Vit.A (IU)	Vit.B1 (mg)	Vit.B2 (mg)	Niacin (mg)	Vit.C (mg)	Ca (mg)	P(mg)	Fe(mg)	K(mg)
Litchi	42	14	87	122	–	trace	0.01	0.31	0.03	–
Loquat	46	–	–	–	–	–	0.03	0.02	0.7	–
Mandarin	49	50	120	60	0.3	68	0.05	0.02	0.1	–
Mango	50	4800	40	50	0.3	13	0.01	0.02	0.3	–
Mangosteen	60	–	–	–	–	–	0.01	0.02	0.02	–
Papaya	40	2020	40	250	0.2	48	0.01	0.01	0.4	–
Passion fruit	93	90	–	–	–	–	0.01	0.06	2.0	–
Peach	38	trace	20	1	0.2	1	0.01	0.03	1.7	–
Pear	47	14	20	30	0.2	trace	0.01	0.01	0.7	–
Persimmon	81	1710	–	–	–	–	0.01	0.01	0.3	–
Pineapple	50	60	–	120	–	63	0.02	0.01	0.9	–
Plum	40	230	120	30	0.3	1	0.02	0.02	0.5	–
Pomegranate	65	–	–	100	–	16	0.01	0.07	0.3	–
Pummelo	45	200	30	–	0.2	20	0.03	0.03	0.1	–
Raspberry	40.2	–	–	–	–	25.5	40.0	44.0	1.00	170
Strawberry	44	–	30	–	0.2	52	0.03	0.03	1.8	–
Sweet orange	53.8	–	–	–	–	50.0	42.0	23.0	0.40	177
Walnut	687	10	450	170	1.6	–	0.10	0.38	4.8	–
Wood apple	97	–	–	170	–	–	0.13	0.11	0.6	–

spacing, cultivar selection, number of plants, and aesthetics. Fruits trees that receive proper care will yield more good-quality fruit than fruit trees that are neglected and left as such for natural growth.

Pre-Planting Care

Selection of Site

Fruit plants are most productive if you carefully match them with the proper climate and planting site. The location should be close to a market and well established fruit growing region to get easily field labourers, and also to get benefit of experience of other fruit growers and to sell the produce through co-operative fruit grower organizations. All fruits require well-drained soil with good water-holding capacity. Access to the necessary inputs like planting material, regular good quality water supply, eco-friendly soil and pest management, live fencing, andcredit or capital for gardening from a local, sustainable source is another important element for successful gardening. Select a site that is free from frost pockets, wet spots, cyclones, frost, hailstorms, strong hot winds and exposure to strong prevailing winds and has ample sunlight for best fruit production. While gooseberries and currants perform adequately well in partial shade, other fruits require direct sun for at least six hours a day, and preferably more. But if production and ease of management are your primary goals, devote separate areas to different fruit trees and small fruits.There must be good transport facilities by road or rail.

Shrubs and weeds should be cleared and site should be ploughed deeply andlevelledbefore planting. In the hilly areas, the land should be divided into terraces depending upon the topography of the land and leveling should be done within the terraces. Terracing protects the land from erosion. If the soil is poor, it would be advisable to grow a green manure crop and plough it *insitu* so as to improve its physical and chemical conditions before planting operations are taken up. Growing marigolds, Sudan grass, or certain mustards (oilseed rape) for a year or two before planting fruit plants can help control certain parasitic nematodes, which are occasionally a problem in certain soils.

Plants like thorny *Agave, Prosopis juliflora, Pithecolobium dulce, Thevetia are* drought resistant, easy to propagate from seed, quick growing, have dense foliage and stand severe pruning and if closely planted in three rows would serve as a good live fencing.

Planting two rows of tall, quick growing, hardy, drought resistant and mechanically strong trees like *Casuarina equisetifolia, Pterospermum acerifolium, Polyalthia longifolia, Eucalyptus globulus, Grevillea robusta, Azadirachta indica* etc as wind breaks is essential in region where strong winds prevail. It is preferable to dig a trench of 90 cm deep at a distance of 3m from the windbreak trees in order to avoid the compelition between the roots of wind breaks and fruit trees for moisture and nutrition.

Because it is difficult to correct nutrient deficiencies and adjust soil pH after planting, it is critical to test your soil before planting to see if you need to add lime and nutrients. Collect subsamples from several locations to provide a representative sample of the site. Send the soil samples to a nearby reliable soil testing laboratory. The soil test will report the pH, the cation exchange capacity and the amounts of various nutrients and soluble salts present. The soil pH is a measure of soil acidity. A pH of 7.0 is neutral; a pH below 7.0 indicates an acid soil, and a pH greater than 7.0 means that the soil is alkaline. The ideal pH range for most fruit crops is 6.0-6.5. Only the blueberry grows best in acidic soils (pH 4.0-5.5). If a soil test shows that your soil is too acidic, you need to add lime (Table 2). Lime will raise the soil pH and supply needed calcium or magnesium. Do not add lime, however, unless a soil test recommends it. An excessively high soil pH can reduce the availability of phosphorus, boron, zinc, iron, and manganese. Reducing soil pH is more difficult. Adding sulphur to the soil will reduce soil pH unless the soil has high carbonate content. If you choose to grow blueberries in very alkaline soils, it is best to remove and replace the soil in the root zone with peat moss or other low pH organic matter. Soluble salts refer to electrolyte compounds in the soil that dissolve in the soil water and are referred to as the salinity level. At high levels soluble salts can reduce water uptake in the plant, restrict root growth, cause root tip burn and in general reduce plant growth and fruit yield. Soluble salts compete with plants for soil water. Seed germination and seedling growth are the most sensitive growth stages to high salinity levels. Plant species and cultivars vary in their sensitivity to soil salts. The salt tolerances of specific fruit crops are given (Table 3). Over fertilization, fertilizer spills or placing fertilizer too close to roots can create soluble salt problems for plants. Correct the soil site problems before planting by adding amendments, organic matter or by planting cover crops about a year before planting.

Table 2: Preferred pH ranges and recommended correction in fruit crops

Fruits	Preferable pH range	High pH correction	Low pH correction
Most fruits	5.5-6.5	pH > 7.0, apply sulphur	pH < 5.5, apply limestone
Blueberries	4.3 - 5.0	pH > 5.5, apply sulphur	pH < 3.5, apply limestone

Table 3: Salt tolerances of fruit crops

Non-tolerant	Slightly tolerant	Moderately tolerant	Tolerant
0 to 2 mmhos	*2 to 4 mmhos*	*4 to 8mmhos*	*8 to 16mmhos*
Blueberry	Apple	Guava	Date palm
Raspberry	Grape	Pineapple	Jojoba
Strawberry	Peach	Jujube	Natal Plum
	Pear	Olive	
	Plum	Jambolon plum	

If planting cannot be done immediately, wrap planting material in sphagnum moss or moist cloth or other material that will prevent them from drying out, and store them at a temperature above freezing. Keep roots moist by spraying with water as needed. If planting will be delayed further, plants should be 'heeled in' by planting temporarily in loose soil as soon as possible.

Laying Out of Orchards

Any method of layout should aim at providing maximum number of trees per hectare, adequate space for proper development of the trees and ensuring convenience in orchard cultural practices. The system of layout can be grouped under two broad categories *viz.* The various layout systems used are the following:

1. ***Square system:*** In this system, trees are planted on each comer of a square whatever may be the planting distance. This is the most commonly followed system, easy to layout and permits inter cropping and cultivation in two directions The central place between four trees may be advantageously used to raise short lived filler trees.

2. ***Rectangular system:*** In this system, trees are planted on each corner of a rectangle. The distance between any two rows is more than the distance between any two trees in a row. The wider row spaces permit easy intercultural and mechanical operations.

3. ***Hexagonal System:*** In this method, the trees are planted in each comer of an equilateral triangle. This way six trees form a hexagon with the seventh tree in the centre. Therefore, this system is also called as 'septule' as a seventh tree is accommodated in the centre of hexagon. This system provides equal spacing but it is difficult to layout and for doing cultural operations. This system accommodates 15% more trees than the square system.

4. ***Diagonal or quincunx system:*** This is the square method but with one more plant in the centre of the square. This will accommodate double the

number of plants, but does not provide equal spacing. The central filler tree chosen should be a short lived one and should be removed after a few years when main trees come to bearing. This system can be followed when the distance between the permanent trees is more than 10m.

5. ***Triangular system:*** Triangular system is based on the principle of isolateral triangle. The distance between any two adjacent trees in a row is equal to the perpendicular distance between any two adjacent rows. When compared to square system, each tree occupies more area and hence it accommodates few trees per hectare than the square system.

6. ***Contour system:*** It is generally followed on the hills with undulated topography to minimize land erosion and to conserve soil moisture. In this system, plants are planted along the contour across the slope. The contour line is so designed and graded in such a way that the flow of water in the irrigation channel becomes slow and thus finds time to penetrate into the soil without causing erosion. In terrace system planting is done on flat strip of land formed along the contours. Terraced fields rise in steps one above the other and help to bring more area into productive use and also to prevent soil erosion. The width of the contour terrace varies according to the nature of the slope. The planting distance under the contour system may not be uniform.

Planting Distance

Planting distances in different fruit crops is decided based on tree growth, kind of fruit trees, rainfall pattern, soil type and soil fertility, vigour of rootstocks, pruning and training system used (Table 4, 5 and 6). If the spacing is too wide the yield per unit area would be greatly reduced. It is profitable to plant the trees closer together and supply the needed water and food materials. If the trees are too close, the trees grow tall rendering pruning, spraying and harvesting difficult. There is root competition for water and nutrients and the trees give lower yield or produce smaller fruits of poor colour. Close planting results in a greater yield per unit area in the early life of the tree but less in later years.

Table 4: Planting distance of fruit crops in different planting systems

Crop	Planting distance(in m)	No. of trees per hectare		
		Square system	Hexagonal system	Triangulars ystem
Mango	10 x 10	100	115	89
Sapota	8 x 8	156	118	139
Acid lime	5 x 5	400	461	357

Hexagonal system of planting accommodates 15% more number of plants while triangular system accommodates 11% lesser number of plants. The calculation of the number of trees per hectare when planted under square or rectangular system is very easy, and is obtained by dividing the total area 'by the area occupied by each tree (a x a in square system or l x b in rectangular system).

Table 5: Fruits and suggested varieties with planting distance

Fruit	Variety	Spacing (m x m)
Aonla	NA-6 , NA-7 (one plant of each variety)	6 x 6
Apple	Golden spur, Red chief, McIntosh, Golden Delicious	4 x 4
Apricot	New Castle, Shipley Early, Turkey, Charmagaz	6 x 6
Banana	Robusta, Dwarf Cavendish, Grand Naine, Rasthali,	2 x 2
Ber	Gola, Seb,GomaKirthi, Umran, Banarasi Karaka	7 x 7
Custard Apple	Balanagar, Mammoth, Atemoya, APK (Ca)-1	5 x 5
Date Palm	Halawy, Khalas, Khuneizi, Barhee	10 x 10
Fig	Capri fig, Smyrna fig, White San Pedro	4 x 4
Grape	Anab-e-Shahi, Thompson Seedless, Arka Vati for South India and Flame Seedless, Perlette, Pusa Urvashi for North India	3 x 3
Guava	ArkaMridula, L-49, ArkaAmulya, ArkaKiran	5 x 5
Karonda	Green, pink, white	2 x 2
Kiwi fruit	Female: Abbott, Bruno, Hayward, MontyMale: Tomuri, Allison, Matua	6 x 6
Lemon	Kagzikalan, Nepali oblong, Lisbon, Eureka,	5 x 5
Lime	Parmalini, VikramandRasraj	5 x 5
Litchi	SwarnaRoopa, Shahi, China	10 x 10
Loquat	Mammoth, Golden Red, Tanaka, California Advance, Pale Yellow,Golden Yellow	5 x 5
Mandarin	Kinnow, Nagpur Mandrin	4.5 x 4.5 and 6 x 6
Mango	Amrapalli, ArkaAruna, Dashehari, Pusa Prathibha, Pusa Arunima, Pusa Surya, Pusa Shreshth, Pusa Lalima, Pusa Peetamber	3 x 3
Mangosteen	-	7 x 7
Mulberry	Black mulberry	7 x 7
Papaya	PusaNanha, Pusa Dwarf	1.25 x 1.25
Passion fruit	Purple,Yellow, Kavery	2 x 2
Peach	Red heaven,Floridasun, Shan-e-Punjab (Low chilling)	4 x 4
Pear	Baghugosha, Punjab Nectar (Low chilling)	5 x 5
Phalsa	Dwarf Type	2 x 2 m
Pineapple	Queen, Mauritius, Kew	30 cm x 60cm\ x 90 cm
Plum	Beauty, Mariposa, Santa Rosa, Kelsey, Kala Amritsari (Low chilling)	4 x 4
Pomegranate	Bhagwa, Mridula, Ganesh, Jyothi	3 x 3
Sapota	Cricket Ball, PKM1, PKM3	8 x 8
Strawberry	Chandlar, Pusa Early Dwarf, Phenomenal	40 cm x 25cm
Sweet orange	BloodRed, Hamlin, Pineapple, Jaffa, Valencia	6 x 6
Barbados cherry	-	2 x 2

Table 6: Planting Density and Expected yields in Fruit crops

Fruit Crop	Planting Distance (m)	Number of plants /acre	Expected yield/plant	Yield /acre
Papaya	1.5 x 1.5	1777	40 kg	70 tonnes
Mango	3 x 3	450	20 kg	12 tonnes
Guava	1 x 2	2000	12 kg	22 tonnes

High Density Planting System

For efficient use of horizontal and vertical space, HDP technologies have been developed in mango, citrus, papaya and guava (Table 6, 7, 8 and 9). Planting of fruit trees rather at a closer spacing than the recommended one using certain special techniques with the sole objective of obtaining maximum productivity per unit area without sacrificing quality is often referred as

'High density planting' or HDP. This technique was first established in apple in Europe during sixties and now majority of the apple orchards in Europe, America, Australia and New Zealand are grown under this system. In this system, four planting densities are recognized for apples viz., low HDP (< 250 trees/ha), moderate HDP (250-500 tree/ha), high HDP (500 to 1250 trees/ha) and ultra-high HDP (>1250 trees/ha). Recently, super high density planting system has been also established in apple orchards with a plant population of 20,000 trees per ha. In some orchards, still closer, planting of apple trees is followed (say 70,000 trees/ha) which is often referred as 'meadow orchards'. Advantages of HDP include early cropping, higher yields for a long time, reduced labour costs and improved fruit quality. The average yields observed in apple is about 5.0 t/ha under normal system of planting and 140.0 t/ha under HDP.

The trees of HDP should have maximum number of fruiting branches and structural branches. The trees are generally trained with a central leader surrounded by nearly horizontal fruiting branches. These branches should be pruned in such a way that each branch casts a minimum amount of shade on other branches. Success of HDP depends upon the control of tree size. This can be achieved by use of dwarfing and intermediate root stocks like MM 106, MM 109, and MM 111 in apple; Quince A, Adam and Quince-C in pears. Use of spur type scions, training and pruning methods can also induce dwarfness.Apple trees trained under spindle bush, dwarf pyramid, cordon systems are suitable for HDP systems. Growth regulators such as diaminozide, ethephon, chlormaquat and paclobutrazal are extensively used to reduce shoot growth by 30 to 0 %. This results in increased flowering in the subsequent years. Tying down the branches to make them grow to an angle of 45° from the main stem are some of the standard practices to control tree size.

The success of HDP depends upon the right choice of planting system. Generally, rectangular planting with single, double and three row plantings are followed. In single row planting, the distance within the row is close, whereas the distance between the rows is wide (4 m x2m). In double row planting, a wider spacing is given after every two rows (4m +2m x2m) whereas in three row planting, a wider spacing is given after every three rows (4 m +2m x2m x2m). In meadow orchard system, a bed of 10 to 15 rows is closely planted (30x45cm) and separated by alleys of 2.5m width between beds. This system is also called bed system.

How far apart you space fruit trees depends in part on how you plan to maintain them. Trees pruned annually are best planted 12 to 18 feet apart. Trees pruned less regularly grow larger and need more space. Espaliered trees, pruned severely several times each year, need substantially less space, in some cases as little as 8 to 10 feet. Your growing conditions also dictate how large trees grow and how much space they need. Trees grow more quickly in very fertile, moist soil, while growth is slow in dry, sandy soils.

Fig. 1: High density orchard of mango variety Amrapali

Table 7: Estimated cost and return from traditional and high density planting in mango

Particular	Traditional system	High density planting
Spacing (m x m)	10 x 10	2.5 x 2.5
No. of plants/ha	100	1,600
Cost of establishing orchard (Rs.)	35,000	75,000
Annual maintenance cost (Rs.)	25,000	40,000
Age of stable yield (year)	8 to 10	7 to 8
Production (kg/ha)	6,000 to 8,000	16,000 to 19,000
Sale of produce (whole sale) @ Rs. 9.0 per kg	54,000 to 72,000	1,44,000 to 1,71,000
Net return (Rs.)	29,000 to 47,000	1,00,000 to 1,30,000

Fig. 2: High density planting in Papaya 6,400 plants/ha (1.25 x1.25 m)

Table 8: Estimated cost and return from traditional and high density planting in Papaya

Particular	Traditional system traditional varieties	High density
Spacing (m)	2.4 x 2.4	1.25 x 1.25
No. of plants/ha	1,736	6,400
Cost of establishing orchard (Rs.)	40,000	75,000
Annual expenditure (Rs.)	25,000	50,000
Age of stable yield (year)	2	2
Production (kg/ha/year)	45,000 to 50,000	80,000 to 90,000
Sale of produce (whole sale) @ Rs. 5 per kg	2,25,000 to 2,50,000	4,00,000 to 4,50,000
Net return (Rs.)	1,60,000 to 1,85,000	2,75,000 to 3,25,000

Fig. 3: High density planting in Guava

Table 9: Estimated cost and return from traditional and high density planting in Guava

Particulars	Traditional system	High density
Rootstock	Allahabad Safeda	PusaSrijan
Spacing (m)	6 m x 6 m	3 m x 3 m
No. of plants/ha	278	1,111
Cost of establishing orchards andmaintenance (Rs.)	35,000	50,000
Annual expenditure (Rs.)	20,000	40,000
Age of stable yield (year)	5	4
Production (kg/ha/year)	8,000 to 12,000	16,000 to 18,000
Sale of produce (whole sale) @ Rs. 8 per kg	64,000 to 96,000	128,000 to 1,44,000
Net return (Rs.)	56,000 to 84,000	1,12,000 to 1,26,000

Fig. 4: High density planting of Kinnow mandarin on troyer citrange rootstock
[Planting density = 3,086 plants/ ha (1.8 m x 1.8 m)]

Table 10: Estimated cost and return from traditional and high density planting in Kinnow

Particulars	Traditional system	High density
Spacing (m)	6 x 6	1.8 x 1.8
No. of plants/ha	278	3,086
Cost of establishing orchards (Rs.)	35,000	75,000
Annual expenditure (Rs.)	25,000	50,000
Age of stable yield (year)	5	4
Production (kg/ha/year)	10,000 to 12,000	22,000 to 25,000
Sale of produce (whole sale) @ Rs. 9 per kg	90,000 to 1,08,000	1,98,000 to 2,25,000
Net return (Rs.)	65,000 to 80,000	1,48,000 to 1,75,000

Selection of Variety

Each kind of fruit tree, even each cultivar (variety), has its own climatic adaptations and limitations. Fruits like banana, pineapple and papaya are suitable for tropical climate, Stone fruits such as peach, sweet cherry, and plum will perform best in the warmer regions of temperate climatic zones (Table 11 and 12). If your place is hot and dry, choose the fruit variety that will best cope with this. Otherwise, when these fruits are grown outside their climatic range, the minimum air temperatures in winter may fall below the survival limit of the tree and spring frosts may kill the blossoms. Many fruit and nut trees from cooler areas have what is called a 'chilling requirement ; the total number of hours needed annually, below 7°C. Without a sufficient chilling period fruit trees may grow well but will simply not set fruit. Hence find out the average chilling hours your area receives and only plant trees that will have their chilling needs met. Chilling hours vary with cultivars of fruit, for example cultivars of apples can be low, medium or high chill cultivars, with between 300 -1200 chilling hours needed.

Most common error commonly made is planting only one fruit tree. It is not only important to grow varieties that taste great and are productive, but you also have to know a little about the need for pollinators when you are choosing a fruit tree. If the fruit variety planted is self-incompatible, flowers and fruits will set but will eventually drop. Most fruit trees grow best when at least two different varieties are planted. The varieties should bloom roughly at the same time and have pollen that is compatible. Check fruit tree catalogs for the varieties that can pollinate each other. Look for self-fertile fruits and varieties if you have limited space.

To determine if a fruit variety will prosper in your area, consult a professional person, an extension or *Kisan* call centre for details (Table 13). If you would like to grow fruit trees that are from a different climatic zone, look for ways to modify the microclimate to improve your chances of success. For example, grapes need dry weather at fruit ripening period. In an area with summer rainfall, try planting grapevines to climb up veranda posts under the shelter of the eaves, to protect the fruit from too much moisture.

Table 11: Fruits for different climates

Tropical climate	Sub-tropical climate	Temperate climate
Mango, Banana, Papaya, Pineapple, Guava, Sapota, Aonla, Custard apple, Carambola, Mangosteen, Jack fruit, Grape, Citrus (mandarin, Sweet orange, Lime)	Mango, Citrus (mandarin, Sweetorange, Lime, Lemon, Grape fruit), Grape, litchi, Avocado, Passion fruit, Ber, Bael, Aonla, Phalsa, Karonda, Pomegranate, Date palm, Strawberry, Loquat, Fig, Kair etc.	Apple, Almond, Walnut, Pear, Peach, Plum, Apricot, strawberry, Cherry, Kiwi fruit, Pecan nut, Pistachio nut and Hazel nut

Table 12: Leading tropical fruit producing states in India

Fruits	States
Mango	Uttar Pradesh, Andhra Pradesh, Bihar, Karnataka, Maharashtra, Orissa, West Bengal and Kerala.
Banana	Maharashtra, Tamil Nadu, Kerala, Andhra Pradesh, Assam, Karnataka, Bihar,Gujarat, Madhya Pradesh and West Bengal.
Citrus	Maharashtra, Andhra Pradesh, Karnataka, Madhya Pradesh, Punjab, West Bengal,Gujarat, Assam and North Eastern region.
Papaya	Maharashtra, Uttar Pradesh, Bihar, Karnataka, Andhra Pradesh, Assam, Orissa andTamil Nadu.
Pineapple	West Bengal, Assam, Kerala, Karnataka, Bihar and Orissa.
Sapota	Karnataka, Gujarat, Maharashtra, Tamil Nadu, Andhra Pradesh and West Bengal.
Grape	Maharashtra, Andhra Pradesh, Karnataka, Punjab and Tamil Nadu.

Table 13: Commercial varieties and new varieties developed in some tropical-subtropical fruits

Fruits	Commercial varieties	New varieties developed/ domesticated using available germplasm
Mango	Alphonso, Dashehari, Totapuri, Neelum, Langra, Baneshan, Kesar	Mallika, Amrapali, Mahmood Bahar, Prabha Shankar, Kesar, ArkaAruna, ArkaPuneet, Manjira, Neeleswari, Neelphanso, Arka Anmol, Sundar Langra, Alfazali, SwarnaJehangir, Neeludin, AU Rumani, PKM-1, PKM-2, Niranjan, Neeleshan, Ratna and Sindhu
Citrus		
a. Mandarin	Nagpur, Coorg, Khasi, Kinnow	Seedless Nagpur Santra and Kinnow
b. Sweet orange	Sathgudi, Mosambi, Blood Red Jaffa Malta, Valencia Late	Valencia Jaffa, Valencia
c. Acid lime	Kagzi lime	Vikram, Pramalini, PKM-1 (Jaidevi), Sel-49 (SaiSarbati)
Banana	Dwarf Cavendish, Robusta, Poovan, Rasthali, Ney Poovan, Monthan	Co. 1 (Coimbatore), Hy. 1 Hy. 2, Gandevi Selection

Contd.

Pineapple	Giant Kew, Queen, Mauritius	Hy. 7, Selection 1/82,2/82
Amla(Aonla)	Banarasi, Chakaiya, HathiJhool (Francis)	Krishna, Kanchan, NA 6, NA7, NA8,NA9, NA10, Agra Bold, Anand 1, 2, 3
Jujube/Ber	Umran, Gola	CHES Godhra, Sel-1
Pomegranate	Jodhpur Red, Muskat, Ganesh, Jyoti, Bassein Seedless	Ganesh, GKVK-1 (Jyoti), G-27
Bael	Mirzapuri	NB 5, NB 9
Sapota	Cricket Ball, Kali Patti	PKM-1, PKM-2, PKM-3, Co. 1, DHS-1,DHS-2
Tropical Grape	Thompson Seedless, Anab-e-shahi, Bangalore Blue, Black Champa, Kishmish Charni,(Sharad Seedless)	ArkaKanchan, Arka Shyam, Arkavati, Arka Hans, Arka Nilmani, Tas-e-Ganesh Manak Chaman, Sonaka, Dilkush
Litchi	China Shahi, Rose Scented Bedana, Calcutta	Suvarna Sudha
Papaya	Ranchi, Coorg Honey Dew	Coorg Honey Dew, Co1, Co2, Co3, Co4,Co5, Co6, CP81, Pant 1,2,3, Pusa Majestic, Pusa Dwarf, Pusa Delicious, Pusa Giant, PusaNanha, Pink Flesh Sweet
Tamarind	Seedling	PKM 1, T-10, Urigam

Preparation for Planting

Do not plant in shady locations or you can expect weak, spindly trees, poor foliage, and poor fruit set. Allow enough room between the planting site and buildings, trees, power lines or other obstructions to allow tree to fill its space when full grown.

Planting time depends on the fruit, climate and area. Most tropical and sub tropical fruit crops like citrus, mango, sapota and guava are planted during spring and rainy season and temperate deciduous fruit crops are planted during dormant period in winter. If the trees are planted early in the rainy season they soon establish themselves and grow vigorously. Care should be taken that planting is done before the growth starts, otherwise trees will suffer severely and will be in poor condition to withstand the next hot weather.

After marking the positions, trees are planted with the help of a planting board which is usually of 15m long, 10 cm wide and 2.5cm thick with a central notch and one hole on either end in a straight line. The planting board is placed in such a way that the tree fits into the central notch. Two small stakes are inserted one in each end hole. The planting board along with the tree marker is then lifted straight up without disturbing the end stakes.

Dig a hole large enough to hold all the root system of the fruit plant and deep enough to cover the roots properly, approximately twice the diameter of the root system, and two feet deep.Fill the holes with a mixture of sand, soil and farm yard manure (1: 1: 1).Add small quantity of insecticide to control any soil borne insects before planting. Irrigate and leave the pits for the soil to settle.

Dip the plant roots in any fungicide before planting. Spread roots out in the hole on the loose soil, ensuring that they are not twisted or crowded.If roots are too long or stiff to easily fit into the hole, trim them slightly. Too much trimming will result in poor growth. If you are planting a budded or grafted plant, care must be taken to see that graft union is 2-3" above the ground level. This is because, above the graft union is the "scion" or actual variety of fruit that you hope to harvest and below the graft union is the rootstock, the roots that take up moisture and fertilizer, anchor the tree, and determine the amount of dwarfing characteristic the fruit tree will have. Improper planting depth will make dwarf and semi-dwarf trees to grow into standard- sized trees. Fill the hole with soil in layers and press down the soil firmly around the roots to insure good soil contact and remove any air pockets.

One common mistake is planting grafted or budded fruit trees too deep. The plants should be planted in such a way that the bud union remains slightly above the ground level. The trees are irrigated soon after planting. Another most common mistake is to put the plants too close together. Allow ample room for growth so you can prune and perform other tasks.

For cross-pollination, plant two compatible trees within 50 feet of one another. If you are not having sufficient space for planting a cross-pollinating variety, consider buying a multi- grafted plant. To save space espalier fruit trees by training them to grow against a wall in a horizontal direction by removing outward growing branches. Leave plenty of room for growing pollinators.

Strong winds can cause damage to the young plants. To protect them from winds, plants should be staked at the time of planting. Some young plants are subject to considerable injury from sunburn. Wrap the trunks with paper or other materials that are not subject to termite attack. Trunks can also be painted with lime.

Intercropping

In young orchards, growing economic crops in inter spaces of the fruit trees during first few years is referred as intercropping. They also act as a cover crop and the land benefits by the cultivation, irrigation, manuring given to the intercrops (Table 14). Water requirements of the intercrops should not clash with those of the main fruit trees. Vegetables are the best inter crops when compared to millets. But whatever may be the intercrop grown, it should be kept well away from the main fruit trees and irrigated independently. The intercropping should be stopped when trees occupy the entire orchard space. Thereafter, green manuring or cover cropping should be only practiced.

In temperate regions peaches are often grown between apple trees. Similarly, in properly spaced mango orchard, guava trees can be planted to bear in two or three years and will produce a number *of* crops before it is necessary to remove them. Such short-lived trees are known as 'fillers'. Papayas, bananas or phalsa may be well grown as fillers in orchards. The fillers should be removed after a few years usually immediately after the main fruit trees have commenced bearing.

Success Story - Intercrops for Disease Control

Among the various diseases that attack mango crop, gummosis is of great economic importance since the trees die within a very short time. The fungus responsible for mango decline is a common soil-borne saprophyte or wound parasite, distributed throughout the tropics and subtropics. The affected trees show symptoms of wilt and die back. About 150 mango trees in village Murar, in Bihar showed abundant gum secretion from branches and main trunk right from the tree base to tree top, wilting, dieback, vascular browning and death of several trees.The observed gummosis in mango trees was often accompanied by damage caused by a new species of trunk borers.The grubs caused severe damage by feeding on the bark inside the trunk, boring upward, making tunnels, thus hindering the transport of water and nutrients from the roots to shoots resulting in wilting and drying of the shoots.Acting as a wounding agent and vector, the trunk borers probably assist in rapid spread of the disease in the orchard. Several chemicals tried to control mango decline showed little or no success.

Turmeric plantation as intercrop in this severely declining mango orchard at village Murar in Bihar was helpful in suppressing the population of trunk borers, termites and gummosis causing pathogens in the soil, and also provided additional income from the harvest of the rhizomes, nine months after planting.Turmeric root exudates or curcumin in rhizomes present in soil probably assisted in disease suppression by reducing the activity and population of trunk borer larvae and soil-borne fungus. The orchard also became free from termite attack after turmeric planting as intercrop in mango. This study indicates that turmeric plantation can be used as intercrop in organic farming systems to control various soil borne pests and diseases in several fruit orchards.

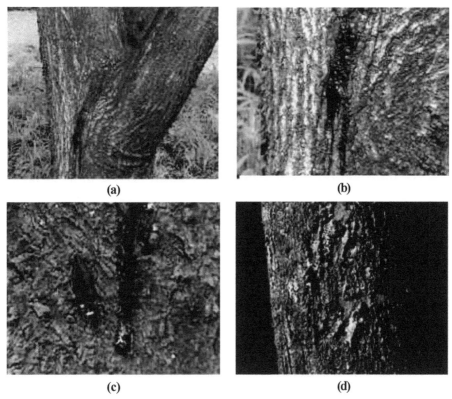

(a)   (b)

(c)   (d)

Fig. 5: Mango decline tree showing gum secretion from **(a)** main trunk right from the tree base, **(b)** middle portion of tree, **(c)** tree top and **(d)** side branches before planting turmeric as intercrop

Fig. 6: Larvae inside the trunk of a declining mango tree

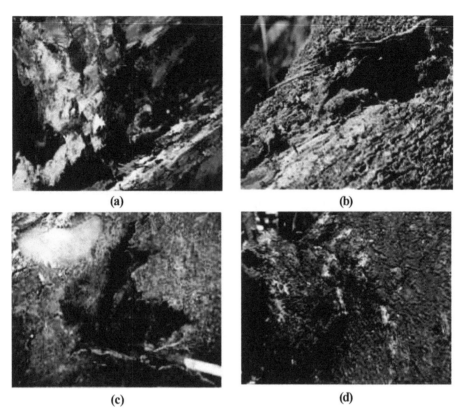

(a)   (b)

(c)   (d)

Fig. 7: Symptoms observed in mango decline tree **(a)** irregular feeding channels, **(b)** webs **(c)** vascular discoloration and **(d)** outer wood showing cracks

Fig. 8: Turmeric plantation as intercrop in mango declining orchard

Fig. 9: Mango trunk showing no symptoms of gummosis after planting turmeric as intercrop in declining mango orchard

Fig. 10: Mango decline tree showing symptoms of dead branches and dry scorched leaves at tree top

Fig. 11: Mango decline tree from control block showing **(a)** Irregular feeding channels in transverse section of trunk, **(b)** feeding channel and vascular discoloration in transverse section of trunk, **(c)** trunk borer below the bark, **(d)** vascular discoloration below the gummosis region.

Table 14: The recommended intercrops for fruit crops

Crop	Age	Intercrop
Mango	Upto 7 years	Leguminous vegetables, Papaya (filler), turmeric
Grapes	Upto 8 months	Snake gourd or bitter gourd in pandal
Apple, pears	Upto 5 years	Potato, Cabbage
Banana	Upto 4 months	Sunhemp, onion

Orchard Cultivation

It refers to the careful management of the orchard soil in such a way that the soil is maintained in a good condition suitable to the needs of the tree with least expenses. This involves maintenance of the physical condition of the soil, its moisture and nutrient content. A good system of orchard cultivation should ensure weed control and saving in moisture and nutrients, very little disturbance to soil and preventing soil erosion and reduced cost of cultivation.

Methods of Soil Management Practices

Clean cultivation: An extensively followed practice in India involves regular ploughingand removal of weeds. It improves aeration. However, there will be loss of humus due to frequent cultivation; injury to the feeding roots; short lived trees; stunted growth; formation of hard pan; soil erosion and depletion of nitrogen. Avoid deep and frequent cultivation and cultivation when the soil is too wet.

Clean culture with cover crops: This involves raising of a cover crop, intercrops or green manure after removing the weeds. Clean cultivation during the rains leads tosoil erosion. It is probably best to plant a green manure crop between the trees early in the rains and plough it into the soil towards the end of monsoon season. In India, green manure crops like Sunhemp, Cowpea, *Dhaincha*, Lupins etc. are more commonly used. Legume cover cropping in grape, mango, guava and other fruit crops is becoming a common practice in the management of orchards. Cowpea and French beans grow well under guava and sapota tree. In some places to prevent soil erosion, certain permanent cover crops like *Calapogonium muconoides, Centrosema pubescens* and *Peuraria phaseoloides* are raised in the alley spaces. Leguminous crops, establish in a short period, dry up during summer to conserve moisture. With summer showers they come up again because of their profuse seeding habit and spread themselves as a vegetative mat by the time the heavy monsoon starts pouring in. Such permanent cover cropping is a common feature in rubber plantations of Kerala and Kanyakumari district.

Mulching: Crop residues like straw, cotton stalks, leaves, saw dust, pine needles, coir dust and other materials like polythene films or certain special kinds of paper are spread in the tree basins and in inter spaces between trees to conserve soil moisture and to control the weed growth. Mulching keeps the soil cool in day; warm at night hours, reduces surface run-off, adds humus to the soil and prevents soil erosion. Fruits are protected and kept clean since they fall on the mulches. It allows the absorption of more rain water andreduces irrigation frequency.However, dry materials used as mulches encourage the risk of fire and consequent damage to trees. Thick mulches may act as places for mice and rodents to live and multiply causing damage to tree trunks and roots by eating the bark and burrowing. The mulching materials should be placed too close to the tree trunk and it should be spread in such a way that they give a good cover to the root system of the trees.

Sod: In this method, permanent cover of grass is raised in the orchard and no tillage is given. This type of orchard cultivation is followed in USA and Europe. This may be useful in sloppy lands for preventing soil erosion. But they compete for soil moisture and available nitrogen and require increased quantity of manure and water. They are harmful to shallow rooted trees. Hence sod may be useful with deep rooted trees because soil moisture will be very low on the top layers.

Sod mulch: This is similar to sod with the only difference is that the vegetation is cut frequently and the cut material is allowed to remain on the ground. This is slightly better than the previous one, as the moisture loss is not great as in sod. In both sod and sod mulch, more nitrogen should be applied to the fruit trees than usual application because the vegetation utilizes more soil nitrogen.

Watering

After the planting hole is filled, irrigate immediately to saturate the soil. After soil is settled, insure that the graft union is still 2 - 3 inches above the soil level. When rainfall is not adequate, irrigate the newly planted trees at least once each week during the first growing season. Earthing up soil around the tree trunks can help prevent spread of diseases like *Phytophthora*.Some trees like mango, citrus require stress period for flowering. Avoid irrigation before flowering period.Irrigate at 10-15 days intervals when fruits are growing till maturity. Stop irrigation during fruit ripening period. Irrigation at the time of fruit ripening period will reduce sweetness.

Weed Control

Weeds can become a serious problem around your trees if not dealt with on a regular basis. Good weed control is very important in the immediate vicinity of transplanted trees to reduce competition. Regular hoeing in the tree basins not only helps in reducing weed growth but will also help in improving soil aeration promoting root growth and growth beneficial microorganisms. Hoeing also helps in killing harmful pests and pathogens by exposing them to sun. Composted mulches are also useful for weed control and for retaining soil moisture.

Training

Training is a relatively new practice in which tree growth is directed into a desired shape and form. To be most productive, fruit plants must be trained to a definite system. There are several training systems. Some fruit trees are trained on single trunk, while some others on Kniffin, vertical trellis, overhead arbour, central-leader or an open centre, etc. Improperly trained fruit trees generally have very upright branch angles, which result in serious limb breakage under a heavy fruit load. This significantly reduces the productivity of the tree and may greatly reduce tree life. Proper tree training also opens up the tree canopy to maximize light penetration. For most deciduous tree fruit, flower buds for the current season's crop are formed the previous summer. Light penetration is essential for flower bud development and optimal fruit set, flavour, and quality. Although a mature tree may be growing in full sun, a very dense canopy may not allow enough light to reach 12 to 18 inches inside the canopy. Opening the tree canopy also permits adequate air movement through the tree, which promotes rapid drying to minimize disease infection and allows thorough pesticide penetration. Additionally, a well-shaped fruit tree is aesthetically pleasing, whether in a landscaped garden, or home garden. Training is satisfactory if they are kept well by regular pruning.

Pruning

Pruning is judicious removal of plant parts to establish balance between vegetative and productive growth. It has strong influence on fruitfulness of plants. In Deciduous fruits, pruning is important to maintain rhythm of fruiting e.g. no pruning in peach means little or no fruiting; in ber, annual pruning is essential; in grapes, annual pruning in north India; two prunings in Maharashtra and three prunings in Tamil Nadu.

In Evergreen fruits, pruning is rarely practiced though it is beneficial in some crops like mango and guava. Allowing trees to grow too tall is another mistake. Pruning trees to manageable size keeps them healthy and is necessary for optimal fruit production and longevity of the tree. While pruning, maintain an adequate number of good leaves for quality fruit production. About 30 to 40 good-sized healthy leaves are needed to produce one good-quality fruit. The safest time to prune most trees is after picking the crop in summer to restrict growth. Winter pruning should be restricted to removal of suckers, dead wood and broken or crossing- over branches to sufficiently open up the canopy of the tree. Pruning cuts should be at an angle so that moisture does not collect and allow for rot and disease to settle. All pruning cuts should be painted with Bordeaux paste to prevent entry of pathogens.

Fertilizing Fruit Plants

Never add fertilizer when planting a fruit tree. The fertilizer can burn the young roots and cause a great deal of damage. A general recommendation is to apply 100-150 grams of 10-10-10 fertilizer or its equivalent 7 to 10 days after planting and the same amount again 40 days after planting. Broadcast the fertilizer evenly, 8 to 12 inches away from the trunk, and work into the top few inches of soil. Trees should be well-watered after fertilizing.In second or third year of planting, addition of 250-300 grams of 10-10-10 in March and again in May will be beneficial. On light, sandy soils, where phosphorus and potassium may be low, twice the amount of 10-10-10 or similar fertilizer should be used. Do not over fertilize, however, because it may result in too much vegetative growth with a loss of yield and quality of fruit or cause injury to the roots and/ foliage. Avoid using fertilizers that contains salt, as this will hinder the growth of tree. Adding organic matter to the soil helps in supplying some plant nutrients; improving soil structure and texture by giving heavier, "hard-to-work" soils a looser texture; and sandy soils to hold more water. If the soil test reveals that organic matter is below 2%, plan to add organic matter into the top 6 inches of the soil before planting. A variety of materials such as leaf manure, vermicompost, cow manure, biodynamic composts etcare available to choose from.

Bioferlizers like *Trichoderma*, Vesicular arbuscularmycorrhiza (VAM), Phosphorus solubilising bacteria (PSB) etcwill not only help in improving the nutrient uptake but will also help in reducing the requirement for chemical fertilizers. *Panchakavya*, an organic solution prepared by adding parts of cow s fresh milk, curd, ghee, urine and dung can also be used as foliar spray on fruit plants to improve plant growth and disease resistance. Fertilizer applications to mature, bearing trees should be based on the growth and vigour of the tree. Since tree s feeder roots is more abundant along the drip line of the tree, fertilizer should be applied along the drip line. Levels of other nutrients needed by fruit crops are best indicated by a leaf analysis during the first growing season.

Pollination and Fruit Set

Fruit trees normally begin to bear fruit soon after they are old enough to flower. Some fruit trees and varieties flower only once in a year, while some other fruit trees flower twice or thrice in a year depending on the climate. Nevertheless, the health of the tree, its environment, its fruiting habits, and the cultural practices you use influence its ability to produce fruit. Adequate pollination is essential to fruit yield. One unfavourable condition may reduce yields or prevent the bearing of any fruit. Some trees like pecans have separate male and female flowers on the same tree. If the male pollen is shed before the female flower is receptive, fruit-set becomes a problem. Some species of fruit trees do not fit conveniently into either category. Pistachios have male trees that produce pollen and female trees that produce fruit. To grow them successfully, it is necessary to plant at least one male tree for every eight female pistachio trees. Most apple trees are self-unfruitful. Plant at least two varieties near one another. 'Golden Delicious', a self-fruitful variety, and 'Jonathan' are the most common pollinators used. Occasionally, fruit trees bear heavily one year and sparsely the next. This is called "biennial bearing." The spring-flowering buds of most hardy fruit trees are formed during the previous spring or summer. Therefore, an especially heavy crop one year may prevent adequate bud formation for the following year or may seriously weaken the tree. Biennial bearing seen in mango, apples etc. is difficult to alter or correct.

Why Fruit Set is Poor in My Orchard?

- Fruitfulness is a condition in which a plant develops flowers, sets fruits and carries them to maturity. Unfruitfulness is a state in which a plant is unable to bloom, set fruit or carry them to maturity. It is a serious problem in many fruit trees and hence it is necessary to understood external and internal causes to obtain acceptable production level.

- External factors are those conditions which impact the life cycle of a plant and include environment before or after flowering e.g. Climate: temperature, chilling requirements, rainfall, cloudy weather, wind and frost; age and vigor of the plant, rootstocks, disturbed soil and water relations, nutrient supply, pruning, insects, pests and diseases, spraying of chemicals during flowering.

- Temperature requirement of plants is quite specific for growth, flowering and fruiting. It profusely influences reproduction, embryonic development and growth of a plant. Optimum temperature ranges from 22-28°C and varies according to fruit species. Deviation in optimum temperature alters flower behavior; variation in day/night temperature and can affect flower initiation and pollination. Extreme temperature fluctuations (high or low) are injurious to flowers and can hinder pollination and fertilization. Temperature above 32°C results in desiccation of stigmatic surface and more rapid deterioration of embryo sac.

- Rainfall at flowering causes damage in several ways- It washes off pollen from anther and stigma, decreases pollen viability, reduces activity of pollinating insects, reduces fruit set or no fruit set. Heavy rains can result in heavy fruit drop. Cloudy weather favors spread of diseases like powdery mildew which usually appear immediately after cloudy weather, delays dehiscence of pollen, adversely effects on pollen germination and growth and has devastating effect on flowers and fruit set.

- Wind is most important agent for transfer of pollen from stamen to pistil. Important in both insect (Entomophilous) and wind (Anemophilous) pollinated plants. Bees and other pollen carrying insects work more effectively in still atmosphere. A reasonable wind speed helps pollination by carrying pollen and is necessary for good fruit set. High wind velocity at flowering hinders movement of insects and affects pollination. Strong and desiccating winds adversely affect fruit setting.

- Frost is a state of environment where temperature of air falls to 0°C or below 0°C. Moisture in air changes into ice and falls on the plant surface; results in irreversible damage to the plant. It may damage flowering or sexual parts of flowers making plants unfruitful.

- Most plants have long juvenile phase and bear only after certain maturity is attained. Young and vigorous trees fail to set fruit. Less vigorous trees set fruit freely.

- Rootstocks have mechanical and physiological influence on scion. Grafted plants bear early crops compared to seedlings and influence fruitfulness e.g. Malling series rootstocks in apple induced precocity; *Troyer citrange* rootstocks in citrus hastens fruit maturity; SohSarkar rootstocks in citrus delays maturity.

- Disturbed soil and plant water relations may lead to abscission of flowers and fruits. During early stages of fruit development, shortage of water is quite harmful. Moisture stress causes disturbance in Carbon- Nitrogen ratio (C/N ratio) and other physiological activities leading to fruit drop. Judicious water supply results in best performance. Abundance of water may cause flower and fruit drop *e.g.* apple and olive.

- Improper pruning, very heavy pruning, and pruning at wrong time can result in poor flower and fruit set.

- Flowers are subject to attack of various insects, pest and diseases; result in serious reduction in fruit set. Insects are both helpful (housefly and honeybee) and harmful (mealy bug, anthracnose and mango hopper) for fruiting. Adopting suitable preventive measures before attack greatly increases fruit set. Useful insects act as pollinators; Plan cultural operation not to harm beneficial insects. Harmful insects like mealy bug suck sap of flowers and developing fruits resulting in drop of flowers and fruits. Hence it is necessary to reduce their threshold level, by adopting suitable preventive and timely control measures.

- Internal factors are evolutionary tendencies (sexuality); sex distribution monoecious, dioecious and pseudo hermaphroditism; structural diversity (floral structure and formation); heterostyly, dichogamy, abortion of pistil or ovules, pollen impotence, genetic sterility and nutritional status of the plant (C/N ratio).

- **Monoecious (Uni-sexual)**: Flowers are imperfect; either staminate or pistillate occurring on same plant, same inflorescence or separate inflorescence; often wind pollinated. Pollination is possible but less likely, ensures out crossing by spatial separation.Common in temperate regions. Few examples are chestnut, hazelnut, walnut, jackfruit, pistachio nut, date palm, papaya, coconut and areca nut.

- **Dioecious (Uni-sexual):** Male and female flowers are borne on separate plants. For proper pollination, fertilization, fruit set and seed development a number of male plants are required to be planted in the vicinity of female plants; otherwise there will be poor fruiting. Commonly found in tropical region. Examples are papaya, date palm, strawberry, Muscatine grape. Japanese persimmon some varieties are monoecious, some dioecious.

- **Psuedohermaphroditism**: Plants produce morphologically perfect flowers having both male and female parts but functionally unisexual (either stamen or pistil is non-functional and behave like either male or female) e.g. grape, plum, pomegranate, persimmon

- **Structural diversity**: Certain minor peculiarities are observed in normal appearing flower either in floral structure or functions preventing self pollination and making cross pollination a rule; absence of cross pollination results in self or cross incompatibility and unfruitfulness. *e.g.,* heterostyly, dichogamy, pollen impotence, abortion of pistil or ovules.

- **Heterostyle:** Morphologically different group of flowers are borne on the same plant. eg Heterostyly: occurrence of flowers with variable style length (common in Prunus fruit). Distyly: short styles, long filaments (thrum type); long styles, short filaments (pin type); (Only compatible mating between pin and thrum type or vice versa); Tri Styly: Stamens and styles having three positions. *e.g.,* long, medium or short; Ex.: Almond, carambola, cashewnut, litchi, pomegranate and sapota.

- **Dichogamy**: In hermaphrodite plants, stamens and pistils mature at different times or receptivity does not coincide with pollen viability/maturity and prevents self-pollination in perfect as well as monoecious flowers.

- **Protandry**: Stamens mature and anthers shed before the pistil is ready to receive pollen (stigma receptivity doesn t coincide with pollen viability and maturity); common in insect pollinated plants, more common than protogyny, ensures out crossing by temporal separation. *E.g.* walnut, coconut, macadamia, custard apple, sapota and passion fruits.

- **Protogyny:** Stigmas are receptive before anthers release pollen. All monoecious plants and majority of dioecious plants are protogynus. *E.g.* banana, plum, pomegranate, avocado, chestnut, pistachionut.

- **Impotence from abortive flowers**: In perfect flowers, reflexed stamens partially defective pollen, or quantity of pollen produced is small, non-viable (e.g. European varieties of Grape) resulting in failure to set and mature fruit. Abortive flowers are more common in plants having indeterminate inflorescence. High pollen viability is necessary to ensure good fruit set and yield.

- **Impotence from degenerated or abortive pistil or ovule**: Any factor that hinders the process of fertilization results in unfruitfulness or sterility. Complete: no flower or no sex organs are formed or fail to attain full development; Partial: either stamens or pistils abortive, sometimes looking normal; Abortion of ovules: *e.g.* mango; multiple ovules and anthers result in both female and male sterility.

- **Genetic incompatibility** means that a plant cannot produce a zygote with its own pollen.

There are two types, sporophytic and gametophytic. If the pollen is of the same allele as that of the stigma, then pollination will not be successful.

- **Poor nutritional status** affects vegetative growth, production of defective pistil and impotent pollen. Excessive N fertilization results in excessive vegetative growth at the cost of flowering and disturbs C/N ratio. Best C/N ratio is CC/NNN.

Fruit Thinning

It is usual for a number of young fruits to drop off during the spring and early summer. This natural thinning is often referred to as the "June drop". With most cultivars, too many fruits will likely still remain on the tree. It is often necessary to remove the excess fruit by hand when it is still very small. Fruit thinning reduces limb breakage; increases fruit size, improves colour and quality of remaining fruit, prevents depletion of the tree and stimulates flower initiation for next year's crop.

Pest Management

Managing disease and insects is big challenge once the trees begin to fruit. Inspect the trees on a regular basis to see if there is fresh damage. Insects like aphids, leaf rollers, mites, moths, slugs, and maggots can all destroy fruit trees. In addition to insects, molds, mildew, blights, scab and brown rot can also be problematic for fruit trees. Sanitation is the key for successful management of insects and diseases. Remove infested or infected leaves, twigs, and fruit from the area and burn them to preventthe source of infection. Dormant oil, applied before buds break, will control scale, mites, and pear psylla. Insecticidal soaps are active against many soft-bodied insect pests. Copper and sulphurcontaining products will control fungal and bacterial diseases.

Because insect infestations and diseases can be very contagious, you must treat any problems quickly to prevent infecting neighbouring trees. Spraying of *Bacillus thuringiensis*, a pathogen capable of attacking several lepidopteron insects has been successfully adopted to control pests on many crops. Using pheromone traps, or sticky traps, breeding sterile insects is another technique for pest control. Many botanicals have the properties to control pests and diseases of plants. Extracts of *neem, dhatura, Calotropis, Pongamia, Callophylum* and custardapple seeds can control a wide range of insects, bacteria, fungi and virus. These pesticides are economical, eco-friendly, easy to prepare andeffective in controlling various pests and diseases.

Controlling Birds

Birds are by far the greatest pests in many fruit trees. Covering the trees with wire cages, plastic netting is perhaps the best method of control. Smaller trees can be wrapped in nets. Use a fine mesh netting to reduce the possibility of birds becoming caught in the net. Individual fruits can also be bagged using butter paper or cloth bags to prevent bird damage, insect and disease attack.Aluminium pie tins suspended by a string or wire above the bushes will twist and turn in the breeze and keep the birds away.

Harvesting

Fruits grow either singly or in clusters depending on the type or variety. Some varieties may be shaken off easily when ripe, others have to be handpicked. Most fruits should be picked when still firm but somewhat green in colour. If left to ripen on the tree, they may turn brown and soft inside. Colour, firmness and flavour are useful indicators in determining when fruits are ready to pick. Some tips are given below.

- Look for change in colourof skin from green to yellow;
- Orseed coat s colourturning brown;
- flesh texture changing from tough and pulpy to crisp and juicy;
- fresh refreshing aromatic smell of ripening fruits and
- Separation of fruit from the tree when picked with an upward motion.

Storage life of harvested fruits is relatively short. Hence, store in refrigerator, or in other cool storage to enjoy fruits for a considerable period of time. One tree can producesufficient fruit to meet the nutritional needs of an average Indian family.

References

Chadha, K. L, 2008. Hand book of Horticulture. ICAR, New Delhi.

Chattopadhyay, T. K, 2010. A Textbook on Pomology, Vol 1: Fundamentals of fruit Growing, Pp. 323.

Das, R. C, 1985. Planning of orchard. (In Fruits of India-Tropical and Subtropical ed.T.K.Bose) Nayaprakash Culcutta.

Gardner, V. R, Bradford, F. C. and Hooker, H. D, 1952. Fundamentals of fruit production. Mc.Graw-hill Book company, Inc.

Hayes, W. B, 1972. Fruit growing in India, Kitab Ghar, Allhabad.

Singh, J. 2002. Basic Horticulture. Kalyani Publication, New Delhi.

2

Frost, Frost Protection and Winter Injury

Ashok Yadav

Division of Fruits and Horticultural Technology, ICAR-IARI, New Delhi

The damage due to frost in fruit crops has been a problem for farmers since starting from their cultivation. Damage caused by frost is a worldwide problem because if all aspects of crop production are well managed, one night of freezing temperatures can lead to complete crop loss. According to Bagdonas*et al.* (1978) frost periods in temperate regions are shorter in duration and occur more frequently whereas in subtropical climates it is occurs with slow moving cold air masses. In tropical regions, there is usually no or very less freezing damage except at higher elevations. In subtropical fruits crops, damages occurs mainly during the winter, whereas in deciduous fruit and nut trees, it occur mainly in spring, but sometimes it occurs in autumn as well. The region prone to frost should adjust the date of planting of crop, microclimate, variety and other preventive measure. According to Attaway (1997) in South Carolina, northern Florida and Georgia, orange fruit were commonly grown before 1835, but losses due to frost damage, farmers did not prefer to grow oranges in that regions. The history of frost damage is more sporadic in nature leading to some major losses from time to time. In December 1990 and 1998

Californiacitrus industry suffered two major damaging events of frost which ranged from little damage in some regions to severe damage in the other region. The economic losses to fruit crops were high in the frost occurring in 1990 compare to frost in 1998.

In 1907 W.C. Scheu, introduced the metal container heaters (*i.e.* stack heaters or smudge pots) for frost protection in Grand Junction, Colorado, USA. The metal container heaters consist of an oil burning device for heating which was more efficient than open fires, later which was known as the HYLO orchard heater. According to Powell and Himelrick (2000) heaters were mainly used to protect crops in orchard. However, in recent history, metal containers for the fire were used to better retain the heat for radiation and convection to the crop. The use of heaters was common practice worldwide for some time in the orchards, but later in USA due to health and environmental problems use of orchard heaters were banned because smoke was terribly polluting the environment. Today at many locations return stack heaters and clean burning propanefuel heaters are legal. Later on the farmers starting to take the crop insurance after the 1990 frost and growers with insurance experienced more damage in 1998. This had resulted because their orchards were more prone to damage or there was less effort to use protection methods because they had insurance.

Frost

The term "frost" is defined as the condition in which an air temperature becomes 0 °C or lower, at a height of 1.25 and 2.0 m above soil level, inside an appropriate weather shelter. The word frost is often confused with the term freeze, but these are having different meanings. The frost is a condition in which water within plants may ormay not freeze, depending on several avoidance factors (e.g. supercooling andconcentration of ice nucleating bacteria) whereas a "freeze" is a condition in extracellular water within the plant changes from liquid to ice (i.e. freezes). Depending on tolerance factors (e.g. solute content of the cells) freeze may or may not result to damage of the plant tissue. A frost event converted into freeze event when extracellular ice forms inside of the plants cell or tissue. Freeze injury occurs due to fall in the value of critical temperature where there is an irreversible physiological condition that leads to death of the plant cells.

1. Radiation/Radiative Frost

A radiation frost characterized by a clear sky and calm wind speed with 5 mph (less than 8 km/h) allow an inversion to develop, low dew point temperatures and air temperatures at night fall below 0 °C but remain above 0 °C during the day. The dew point temperature is the temperature reached when the air is cooled until it reaches 100 per cent relative humidity, and it is a direct measure of the water vapour content of the air. In radiative frost, heat is radiated from the earth's surface to outer space i.e. sky.

Type of Radiation Frost: It is mainly of two types.

- *White frost/Hoar frost :* It occurs when atmospheric moisture deposits onto the surface and forms a white coating of ice. It causes less damage than the black frost.

- *Black frost:* This type of frost occurs when temperature falls below 0 °C and resultsfew or no ice crystals formation because the air in the lower atmosphere is too dry. The formation of ice crystals depends on the dew point or frost point. The black frost causes more damage than the hoar frost.

2. Advective Frost

Advective frost are characterized by cold, dry winds and, It is characterized by cold air masses (450-3000 feet thick) with cloudy conditions having moderate to strong winds, low humidity, no temperature inversion and in many cases, the dew point is below freezing. It occurs mainly when a cold air mass moves into an area and replace the warmer air and resulting into freezing temperatures. Often temperatures will drop below the melting point (0°C) and will stay there all day. Protection success limited in the advective type of frost

Table 1: Comparison between radiation and advective frost

Type of Frost Parameters	Radiation Frost	Advective Frost
Sky	Clear	Cloudy
Wind Speed	Less than 5 MPH	Higher than 5 MPH
Thickness of cold air mass	30-200 feet (9-60 m)	450-3000 feet thick
Temperature Inversion	Inversion occurs	No inversion
Protection	Easy to protect	Difficult to protect
Humidity	High (Hoar frost) & low (Black frost)	Low

How Does Frost Occur?

In orchard during the day the soil heats up and becomeswarmer and at night heat is lost from the soil and to some extent from the fruit trees and vines. A frost results due to fall in the temperature to 0°C, at ground level. This fall/decrease in temperature mainly depends on the following things:

(i)   Amount of heat stored in the soil during the day

(ii)   Amount of heat lost by radiation at night

(iii) Flow of heat form the soil to surface or plant surface

(iv)   Moisture content of the air.

Loss of heat through radiation is maximum on a clear night, while clouds have a blanketing effect and wind will mix the air layers bringing warmer air down. The conversion of water vapour into water (dew) results heat and when the temperature cools at night, the temperature of the air in contact with the plants and soiled crease to below the "dew point' resulting moisture to condense and form dew. This gives off heat and retards the temperature drop. When the temperature starts to fall at 0°C it freezes and heat is released as dew changes to frost and ifdecrease in temperature continues the water in the plant cells freezes and ruptures the cell walls with a characteristic burnt appearance of the plant. The risk of frost increased into a condition when combination of dry soils, more clear spring days and nights occurs significantly.

Methods for Frost Protection

1. *Passive protection*
Passive methods should be used in advance of the actual freeze danger. These are the most economical and effective methods and are widely used.

A. *Biological Methods*
- Plant selection and genetic improvement
- Modifying plant genetics by inducingfrost resistance
- Seed treatment with chemicals
- Adjustment in the planting dates (annual crops) & flowering (Perennial fruit crop)
- Growth regulators and other chemical substances

B. *Ecological Methods*
- Site selection
- Modification of the landscape and microclimate
- Nutritional management
- Soil management
- Cover crop (weed) control and mulches

2. Active protection
Active protection should be used immediately before and during the occurrence of the frost. It is unable to prevent the advective frost and is only effective for radiative frost conditions. For active freeze protection good forecasts of minimum temperature and wind conditions of orchards is very useful. Active protection is mainly depend on the reduction of heat loss from the soil and plant surface, mixing of the air to break up the temperature inversion, or by increasing the heat of surrounding to maintain the temperature above the danger point.

Following are the Active Protection Described
1. *Covers*
- Covers without supports
- Covers with supports

2. *Water*
- Under plant sprinklers
- Surface irrigation

- Over plant sprinklers
- Micro sprinklers
- Artificial fog

3. Heaters

- Solid fuel
- Liquid fuel
- Propane

4. Wind machines

- Horizontal
- Vertical
- Helicopters

5. Combinations

- Fans and heaters
- Fans and water

Table 2: Critical temperature (T_C; °C) values for several deciduous fruit tree crops

Crop	Stage	10% kill	90% kill
Apples	Silver tip	-11.9	-17.6
	Green tip	-7.5	-15.7
	1/2" green	-5.6	-11.7
	Tight cluster	-3.9	-7.9
	First pink	-2.8	-5.9
	Full pink	-2.7	-4.6
	First bloom	-2.3	-3.9
	Full bloom	-2.9	-4.7
	Post bloom	-1.9	-3.0
Apricots	Tip separates	-4.3	-14.1
	Red calyx	-6.2	-13.8
	First white	-4.9	-10.3
	First bloom	-4.3	-10.1
	Full bloom	-2.9	-6.4
	In shuck	-2.6	-4.7
	Green fruit	-2.3	-3.3
Cherries (Bing)	First swell	-11.1	-17.2
	Side green	-5.8	-13.4
	Green tip	-3.7	-10.3
	Tight cluster	-3.1	-7.9

Contd.

	Open cluster	-2.7	-6.2
	First white	-2.7	-4.9
	First bloom	-2.8	-4.1
	Full bloom	-2.4	-3.9
	Post bloom	-2.2	-3.6
Peaches (Elberta)	First swell	-7.4	-17.9
	Caylx green	-6.1	-15.7
	Caylx red	-4.8	-14.2
	First pink	-4.1	-9.2
	First bloom	-3.3	-5.9
	Late bloom	-2.7	-4.9
	Post bloom	-2.5	-3.9
Pears (Bartlett)	Scales separate	-8.6	-17.7
	Blossom buds exposed	-7.3	-15.4
	Tight cluster	-5.1	-12.6
	First white	-4.3	-9.4
	Full white	-3.1	-6.4
	First bloom	-3.2	-6.9
	Full bloom	-2.7	-4.9
	Post bloom	-2.7	-4.0
Prunes (Italian)	First swell	-11.1	-17.2
	Side white	-8.9	-16.9
	Tip green	-8.1	-14.8
	Tight cluster	-5.4	-11.7
	First white	-4.0	-7.9
	First bloom	-4.3	-8.2
	Full bloom	-3.1	-6.0
	Post bloom	-2.6	-4.3
Grape	New growth	-	-1.1
	Woody vine	*-20.6*	-
	French hybrids	*-22.2*	-23.3
	American	-	-27

[*Source:* Proebsting and Mills, 1978; Krewer, 1988]

Table 3: Susceptibility of fresh fruits to freezing injury

Severity	Crops
Most Susceptible	Apricots, Avocados, Bananas, Berries (except cranberries), Lemons, Limes, Peaches, Plums
Moderately Susceptible	Apples, Cranberries, Grapefruit, Grapes, Oranges, Pears
Least Susceptible	Dates

[*Source:* Wang and Wallace, 2003]

Table 4: The highest freezing temperature for fresh fruits

Common Name		Scientific Name		Temperature
Acerola; Barbados cherry		*Malpighiaglabra*		-1.4
Apple		*Maluspumila*		-1.5
Apricot		*Prunusarmeniaca*		-1.1
Asian pear, Nashi		*Pyrusserotina; P. pyrifolia*		-1.6
Avocado	cv. Fuerte, Hass			-1.6
	cv. Fuchs, Pollock			-0.9
	cv. Lula, Booth			-0.9
Banana		*Musa paradisiaca* var. *sapientum*		-0.8
Barbados cherry		*Malpighiaglabra*		-1.4
Berries	Blackberries	*Rubus* spp.		-0.8
	Blueberries	*Vacciniumcorymbosum*		-1.3
	Cranberry	*Vacciniummacrocarpon*		-0.9
	Dewberry	*Rubus* spp.		-1.3
	Elderberry	*Sambucus* spp.		-1.1
	Loganberry	*Rubus* spp.		-1.7
	Raspberries	*Rubusidaeus*		-0.9
	Strawberry	*Fragaria* spp.		-0.8
Cactus pear, prickly pear fruit		*Opuntia* spp.		-1.8
Carambola, Starfruit		*Averrhoacarambola*		-1.2
Cherimoya; custard apple		*Annonacherimola*		-2.2
Sour Cherry,		*Prunuscerasus*		-1.7
sweet Cherry,		*Prunusavium*		-2.1
Citrus	Calamondin orange	*Citrus reticulata x Fortunella* spp.		-2.0
	Lemon	*Citrus limon*		-1.4
	Lime, Mexican,	*Citrus aurantifolia;*		-1.6
	Seville; sour	*Citrus aurantium*		-0.8
	Pummelo	*Citrus grandis*		-1.6
	Tangelo, Minneola	*Citrus reticulata x paradisi*		-0.9
	Tangerine	*Citrus reticulata*		-1.1
Coconut		*Cocosnucifera*		-0.9
Currants		*Ribes* spp.		-1.0
Date		*Phoenix dactylifera*		-15.7
Fig		*Ficuscarica*		-2.4
Garlic bulb		*Allium sativum*		-2.0
Gooseberry		*Ribesgrossularia*		-1.1
Grape		*Vitisvinifera*	*fruit*	-2.7
			stem	-2.0
Grape, American		*Vitislabrusca*		-1.4
Jujube; Chinese date		*Ziziphusjujuba*		-1.6
Kiwifruit;		*Actinidiachinensis*		-0.9
Longan		*Dimocarpuslongan*		-2.4
Loquat		*Eriobotrya japonica*		-1.9
Mango		*Mangiferaindica*		-1.4
Mushrooms		*Agaricus,* other *genera*		-0.9
Nectarine		*Prunuspersica*		-0.9

Contd.

Olives, fresh		*Oleaeuropea*	-1.4
Papaya		*Carica papaya*	-0.9
Peach		*Prunuspersica*	-0.9
Pear, European		*Pyruscommunis*	-1.7
Persimmon, kaki		*Diospyros kaki*	
	Fuyu		-2.2
	Hachiya		-2.2
Pineapple		*Ananascomosus*	-1.1
Plantain		*Musa paradisiaca* var. *paradisiaca*	-0.8
Plums and Prunes		*Prunusdomestica*	-0.8
Pomegranate		*Punica granatum*	-3.0
Quince		*Cydoniaoblonga*	-2.0
Sapotes			
	Caimito, star apple	*Chrysophyllumcainito*	-1.2
	Canistel, eggfruit	*Pouteriacampechiana*	-1.8
	Black sapote	*Diospyrosebenaster*	-2.3
	White sapote	*Casimiroaedulis*	-2.0
Tamarind		*Tamarindusindica*	-3.7

[*Source:* Whiteman, 1957]

Winter Injury

Winter injury consist of a myriad of plant symptoms and damage which mainly occur when cold temperatures reach to a critical level resulting into damage of roots, stems, bud tissue, wind scald and sunscald of leaves and bark, bark splitting, drooping and rolling of leaves, and limb breakage. It is most severe in the upper region of the root system, lower trunk and the crown region. The damage in these region leads to splitting of the cambium layer which affect the flow of nutrients and can increase the incidence of wood rots or insect damage. The affected tree will often leaf out in spring season, since energy is stored in the buds themselves, but if that energy is utilized, the tree is unable to pull more resources from the roots and the affected areas will lead to die back. Winter damage is although phenomena for most trees, if it occurs continuously on tree, will leads to weaken the trees by infecting with insect pests and disease.

Occurrence

Winter injury mainly occurs when temperatures falls below the critical level that each species can tolerate. Usually wood is more cold tolerant than flower buds. Tree trunks and branch crotches are the slowest to harden off, and the most vulnerable to cold temperatures. During late fall or early winter, when cold temperatures occurs it results in winter injury and when temperatures become warm or swing erratically in short periods of time it also results in winter injury. Winter injury results in death of the tree immediately, loss of yields and causes a shorter tree life by making the tree more susceptible to pests and disease.

Symptoms

The symptoms of the winter injury depends on the severity of the various factors such as length of exposure to cold temperature, inherent plant hardiness, plant vigor, moisture in plant part, soil moisture, drainage, location, variability in cultivar and different plant part viz. stem, leaves, buds, flower buds etc. The symptoms of winter injury vary from crop to crop and are more severe in colder areas. The damage will be more on the tree as the stress increases with moisture deficits and hot temperature during the winter season leading to collapsing of plants or poor fruit set. The following symptom of winter injury will occurs in the trees:

- Weaker growth in older trees
- Killing of vegetative and flower bud
- Dieback of the trees after tip damage
- Grape bud, cane and trunk damage some vines are killed to the ground
- Slow leaf out in the tender apple cultivars like Jonagold and Mutsu
- Development of cold pockets with dead branches in young apple trees
- Day neutral strawberries with weak growth and blackened crowns
- Tip killing in raspberries and blueberries
- Desiccation of the affected tissue or plant part
- Splitting of bark and sap exudation from limb and trunk area
- Wilting, water soaking and browning of the leaf and wood
- Dead leaves persistently hanging on stems and in severe condition leaf drop also occurs
- Discoloration of twig, limb and trunk

Types of Winter Injury

Winter injury on fruit trees can take many forms, which are described as below:

Blackheart

It is fairly common type of winter injury in which gumming occurs which leads to cell death and also pith is killed followed by darkening of heartwood. In later stages it results into weakening of the trunk and branches, but its recovery will be rapid in healthy trees. Nursery and young trees are mostly affected with black heart. It is mainly found in apple, peach, pear, plum and cherry fruit crops.

Cambium Injury

The severity of cambium injury is increased by warm temperatures followed by cold temperature resulting into hardening off the cambium. This leads to secondary infection of canker and fungi resulting in weakening of the plant. Often gummosis may also occurs but not always. Cambium injury is mostly found in stone fruits such as peaches, plums, apricots, and cherries.

Collar/ Crown Injury

It results into winter killing of bark near the ground and hardening of may be late in the collar injury. It most commonly found in the apple cultivar Northern Spy and Gravenstein.

Crotch Injury

The limbs which are upright with narrow crotches angles are more prone to be winter injury. This type of injury may occur on up and down of the limb. So to avoid the crotch injury the crotches angle should be more and branches must be in horizontal direction.

Desiccation Injury

The desiccation injury results into browning, yellowing and purple discoloration of evergreens in the winter or early spring. Desiccation is occurred when winds in winter remove the moisture in leaf and soil moisture is also frozen, which results into unavailability of water. The plants with broadleaved are more vulnerable to desiccation injury than the narrow leaved plants. The curling of leaf and browning of leaf margin is mainly due to desiccation injury. It can be prevented by foliar spray of chemical anti-transparent (to prevent moisture loss) and mulching with leaves and wood chips around the base of plant to avoid freezing of the soil moisture.

Winter Sunscald

Winter sunscald is also known as southwest injury because trunks and branches of the tree facing south and southwest direction warm the most in that side they get the most direct sun later in the day when air temperatures are warmest. The damage in winter scald occurs when bark of the tree warms up on sunny days and within the plant, previously dormant cells become active in response to the warmth. The injury occurs by killing of active cells within the trunk and limbs of sensitive trees during winter months. When temperatures fall below freezing point during the night time, the newly activated cells lose some of their cold-hardiness which will lead to cracking and discoloration of the bark. It mainly occurs in peach and apple which can be overcome by white latex paint.

Trunk Splitting

Trunk splitting mainly occurs in late fall/early winter with a rapid decrease in temperature. Splitting/cracking may or may not close up and heal over. It can extend all the way to the pith and common occurring in apple and sweet cherry. It also behaves as "the beginning of the end" with stone fruit.

Dieback/Shoot Death

When plant has not fully acclimated/hardened off, tip of the plant is damaged by cold temperature leading to shoot death which when further proceeds and cause dieback. It s mainly occurred in young trees which are similar to a heading cut. Nitrogen fertilization and late pruning should be done properly to overcome dieback problem.

Heaving Injury

Heaving injury is found plants which are mainly grown in heavy clay and wet soils. It occurs by freezing and thawing of soil alternatively which will force shallow rooted plants to come. To overcome rapid fluctuation in soil temperature and heaving, applications of mulching before the soil freezes in late fall.

Root Death

Winter injury is affecting more on aboveground portion rather than the underground portion of the tree. It is common phenomena in apple in which roots are damaged/ killed at -4°C to -12°C.

Frost Crack

The south side of a deciduous tree trunk during sunny winter days is warmed and the tissues expand. The sudden drop in temperature after sunset makes the bark to cool and contract faster than wood resulting into splitting of bark. The process of repeated heating and cooling leads to splitting in trunk.

Low Temperature Injury

The temperature near zero or sub-zero will be injurious to fruit plant if it remains for longer duration and may kill the buds, twigs and branches of non-vigorous trees and shrubs. Plants which are lacking in appropriate soil and tissue moisture are very susceptible to low temperature injury. Defoliation during the growing season and pruning untimely may leads to low temperature injury. Low temperature injury is non-reversible process because once it s occurred it cannot be reversed. It commonly found in peach, apricot, cherry and nectarine fruit crops.

Snow and Ice Injury

The structural damage to fruit trees and shrubs occurs when heavy loads of snow and ice combined with wind. When limb breakage occurs, immediately pruning of all the damaged branches to sound woods and when accumulation of ice takes place, it should not be removed because limbs are very brittle and removal of ice may cause more severe damage to the tree.

Late Spring Frosts

In late spring frost if new leaves buds are damaged, most of the plants will quickly do budding except for some evergreen trees and frosted flower buds are usually lost for that particular season. Frost results in formation of tiny new leaves to cup, twist, curl, crinkle or appear shot full of holes. The affected leaves results in increase in size, which appear tattered, or as if chewed by some insects. This type of injury does not lasting damage to the plants.

Mitigation Strategies

To reduce or eliminate loss of winter injury mitigation strategies need to adopt which are as follows:

- Production insurance is purchased before winter.
- Selecting sites less susceptible to cold.
- Thinning of the hedgerows.
- Selection of hardy cultivars and rootstocks for the local climate and soil conditions.
- Restrict soil applied nitrogen to early season.
- Application of loose organic mulch on the root zone by maintaining soil moisture.
- Pruning, fertilization or irrigation in late in the season should be avoided because it might stimulate new growth.
- Pruning of dead and severely damaged wood.
- Provide shading or a windbreak during the winter months
- Paint tree trunks with white latex paint
- Use of wind machines.
- Properly and timely irrigation throughout the spring, summer and fall.
- Removal of as much fruit as possible on damaged trees allowing the tree to recuperate rather.

- Use of cold frames, greenhouses, fire, smudge pots, incandescent lights and aerial watering.

Winter Injury in Relation to Specific Fruit

Apple

The major hindrance for fruit crop production is winter injury. The MN hardy apple trees in the middle of winter can tolerate very cold temperatures. After the dormancy breaks the swelling of the buds starts, after which they lose the ability to withstand in cold winter temperatures. At initial stages swollen buds withstand cold temperatures without any damage but after bud opening temperature at near 20°C cause harm to some buds. Winter damage just after the transition out of dormancy may result in dead flower buds, twigs, shoots and branches. Low temperature injury in the dwarf and semi-dwarf rootstocks is more vulnerable in young trees due to the late growth resulted from rainfall and high temperature late in the season and premature dormancy due to the lack of moisture and available water in the soil. According to Wildung*et al.* (1973) apple roots have less ability to acclimatize to low temperatures and root can be damaged by temperatures below –8°C (Embree, 1988). In Northern parts of the Canada and United States, cold temperature injury to the root of apple trees can resulted in reduction of yield (Czyncsyk, 1979). Root injury symptom become evident the next spring when trees „„leaf out followed by wilting and death of new growth in severe cases and stunted shoot growth in less severe cases. Embree, (1988) in his studies found that Malling rootstocks that are planted in orchards of the United States are lacking cold-hardiness or tolerance of subfreezing temperatures whereas B-9 rootstock showed greater cold-hardiness (Quamme and Brownlee, 1997). Various attempts have been made to quantify the relationship between winter injury and weather variables (temperature) and Blackburn (1984) gave criteria for determining the susceptibility of the apple trees to winter injury based upon minimum monthly temperature and a calculated rate of air-temperature change. According to Phillips (1990) "winterkill day" is a day with less than 4cm snow cover and with minimum temperature below -15°C (moderate) and -20°C (severe). Buszard (1981) had observed that early cultivars were less affected by the winter injury because early harvest of the trees helped them to start their dormancy earlier than late cultivars while Quamme *et al.* (2010) had found that apple trees are more prone to fall freeze events during the acclimation stage.

Banana

The edible banana plants don t perform well in regions having freezing temperatures, whereas the ornamental banana plants are a somewhat more cold tolerant. The extent of damage in banana depends on the low temperature, time duration exposed to freezing temperatures. Banana plant is highly susceptible to winter injury and produce edible fruit only without exposure to freezing temperatures. During ripening process if light frost occurs, it will stop the ripening whereas a hard freeze will permanently stop the ripening process by killing the fruit which later turn into black and mushy. Leaves show burning symptom when light frost occurs, but the damaged plant will produce new leaves to replace the leaves (Figure 1.). The damaged leaves should be cut off when new leaves appear. The pseudostem of banana turns brown and oozy when it exposed to a hard freeze conditions. To overcome this problem cut the pseudostemupto ground level and wait till warmer temperatures occurs and new shoots will sprout from the rhizome. During the winter if prolonged freezing temperatures prevail, it will kill the roots below the groundsurface which will later on kill all the top portion of the banana plant. If some part of the roots survives, it will resprout in the spring. The *Musa basjoo* (Japanese fiber banana) is the hardiest banana plant which can withstand- 20 °F.

Fig. 1: Banana leaves damaged due to frost injury

Blueberry

In blueberry crop, winter injury is one of the limiting factors for its growth and development. Young plants are significantly more prone to damage than mature plants of ten year age or more. Branch tips are most susceptible to winter injury in blueberry. Mild winter injury results in only a few scattered dead branch tips and also reduces the amount of fruit produced whereas in severe winter injury the plants will die. Vigorous stems (one or two feet height during the summer) are more susceptible to winter injury than slow growing stems. The branches susceptible to winter injury should be cut back to living tissue. In blueberries avoid fertilizing with nitrogen after the middle of June. In late summer, if the plants are supplied nitrogen the branches will keep growing into the fall. During dormancy provision of sufficient moisture should be there with irrigation or rainfall. Winter injury can also be minimized by planting blueberries in areas that are protected from winds during the winter. By application of mulching, young plants can be protected from winter injury, snow or a breathable row cover. Row covers should be put in November over the berries and removed in few days after the snow melts. Straw mulch should do carefully because it attracts mice which then can feed on the blueberry bark.

Citrus

Citrus trees are evergreen in nature never go to dormant stage except Trifoliate orange. Due to broad leaves of citrus these trees are more prone to cold damage as soon as freezing temperatures come around the tree. The trees having lesser than five years old are more prone to frost damage compare to older trees. To prevent citrus from cold damage the selection of site should be free from frost or cold winter along preventive measures will help to overcome the problem. Most of the citrus varieties perform well where the average annual minimum temperature remains above 25 to 30 °F. The hardiest in citrus is the ornamental trifoliate orange, which produces inedible fruit and perform well at 5 °F whereas grapefruit, Navel orange, tangerine and sour orange perform well at temperature down to 20 °F without any severe damage. The least cold hardy species i.e. lemon and citron show symptoms of injury, when temperatures fall below 25 °F. To protect from the winter injury planting of citrus tree should be on the south side having full sunlight which will give protection frost damage. Pruning and fertilizers in the fall or winter should be avoided because it will encourage new growth of flushes which is most susceptible to cold damage. Covering of small trees with burlap sacks or blankets forshort term will also provide protection from frost. Freeze damaged leaves of citrus tree become hard and brittle showing wilting and discoloration after thawing. If damage is not severe, plant recovers the leaves but discoloration remains as such. The severely damaged citrus leaves collapse and dry out, hanging on the tree for several weeks (Figure 2). When symptom occurs on twig and branch, it will appear as darkened discoloration, splitting and patches of dead bark. When wood of citrus is severely damaged, leaves remain for longer time on the tree whereas if the wood is damaged up to some extent the leaves fall off from the tree rapidly.

Fig 2: Severe frost injured citrus tree

Grape

Winter injury cause direct losses in grape production and value added product *i.e.* wine production. For example, at Finger Lakes region in New York winter resulted in direct crop losses of $5.7 million and a value-added estimate of lost wine sales of $41.5 million injury from a single event in January 2004 whereas total losses in the wine industry from that one freeze were estimated at $63.6 million (Martinson and White, 2004). During the dormant grapevine s period, tissues which survive severe freezes show a light green or creamy green color when cut. The membranes surrounding each cell and around the sub-cellular organelles are destroyed during freeze injury. Winter injury in the phloem tissue of cordons and trunks starts and progresses from the outer to the inner phloem. The severe effects of winter injury to grape vines is on the primary bud because it is most sensitive to winter injury and its failure to emerge is a indicator that the vine has been winter-injured. The complete failure of bud break (blind nodes) indicates that injury has occurred into the secondary and tertiary buds. Severe winter injury of buds is often associated with injury to the vascular system of canes, cordons and trunks. Therefore, the shoots which survive will not be able to draw sufficient water and nutrients to maintain transpiration water. When the vascular system fails to support shoot and crop development it may cause sudden wilting and death of vine s canopy in late spring or summer.The catastrophic injury on trunks of grape vines from winter injury is difficult to recover whereas partially killed grape vine and localized injury of canes and cordons may be manageable with judicious training and pruning. The hidden buds at the bases of canes or older woody branches will often break dormancy and push as rapidly growing shoots after severe winter injury to above-ground parts of the vine. Similarly, when winter injury kills the entire canopy, the viable buds of base of the grape vine, which were dormant for long time, may erupt as very vigorous suckers.

Mango

Mango trees are the most cold sensitive fruit plants and perform best in subtropical and tropical climates. When temperature goes below 30 °F, leaves and twigs of the plant suffer from serious damages. The high temperature cause less damage to the plant compare to the low temperatures. Frost damage occurs when ice forms inside the plant tissue and injures plant cells. Freeze damage becomes more severe when it combined with cold winds (Figure 3).

Fig. 3: Frost injured mango branches

Olive

In olive the major damage were defoliation, bark split, and limb dieback. In the freeze damaged tissue leaf chlorosis and olive knot infestations is most commonly occurred symptom. The freeze killed many fruiting buds and many fruit that were set developed slowly or poorly in the absence of leaves. The overall economic impact as a result of the freeze is unclear. Excessive irrigation, especially when continued after harvest, may elicit more vegetative growth and may make trees more subject to damage. Conversely, limited water may promote tree desiccation, thereby increasing hardening off as well as retarding vegetative growth (Figure 4).

Fig. 4: Olive tree injured with frost

Papaya

Among tropical fruit crops, papaya (*Carica papaya*) does not perform well in winter and a freeze conditions will cause severe damage/death to the entire plant. The fruit are sensitive to the cold, both on the tree and in storage. The papaya plant can be protected from winter injury by providing irrigation to the roots. The roots that remain submerged for 48 hours will kill the plant, no matter what the temperature is outside. The temperature below 32 °F will leads to minor damage to plants, but long term exposure to subfreezing temperatures can kill the entire papaya. Winter injury to the papaya plant varied from killing of the growing tips, foliage, fruit and entire plant (Figure 5). Cold weather below 60 °F slows the ripening process of the fruit and growth of the plant. Whereas low temperature occurs after fruit formation then it will affect taste, flavour and quality of the fruit.

Fig. 5: Papaya frost injury

Pear

In pear orchards it was observed that Bartlett, Anjou, and Bosc showed severe injury/death in trunks and crotches whereas Flemish Beauty, Cornice and Easter varieties showed much less injury showed only slight injury. Whitewashing of pear tree trunks and crotches was found effective in reducing injury due to direct exposure to the afternoon sun's rays following low night temperatures during the winter. Board shields were also effective.

Strawberry

Strawberry plants are considered as tender and herbaceous perennial with fairly susceptible to low winter temperatures therefore require additional protection to survive the extreme winter temperatures. When the crown temperature drops from - 12°C to -20°C, winter injury in strawberry crowns occur and this injury will lead to browning of crown tissue in the spring but this will not affect overall yield of strawberry plant. There is stunting of leaves, decrease in leaf numbers, formation of fewer flowers and fruit, when there is fall in crown core temperature from-6°C to -9°C. Straw mulching protects crowns and shallow root systems of the strawberry crop from low temperature injury. By keeping soil temperatures more uniform will protect strawberry from drying out from cold and dry winds.There are many factors responsible for cold tolerance of strawberry which differs at varietal, species and genera level. Yao *et al.* (2009) conducted a research at Minnesotta on strawberry, where nine varieties were grown under mulch or with no mulch for two seasons. They found that mulched strawberry plants were less injured compared to not mulched plants. A study conducted at Canada by Gagnon *et al.* (1990) resulted that the June-bearer cultivar Redcoat and day-neutral cultivar Tristar were killed at 19 °F compared to 20.8 °F for the day-neutral „ Hecker. The study conducted on day-neutrals in Quebec indicated that fall fruiting of day-neutral strawberry varieties may also affect the cold tolerance by reducing the accumulation of nitrogen, total non-structural carbohydrate and starch level in the plants. Fall fruiting also reduced the cold tolerance and plants that were de-blossomed were killed at 21.6 °F, whereas plants that fruited continuously were killed at only 23.2 °F (Figure 6).

Fig. 6: Strawberry fruits affected with frost injury

Walnut

In walnut the winter injury mainly occurs just before the dormancy period which occurs during the early fall frost. Young walnut trees are more prone to winter injury than the mature walnut tree because mature plant enters into dormancy due to which they escape from the winter injury. The foliage which is green in colour is more susceptible to injury and the frosted leaves drops prematurely exposing the young twigs fully exposed to sun light. These exposed twigs don t show any symptom of frost but prolonged exposure to sunlight leads to sun burning symptoms in the mid of January. The oblique or horizontal branches commonly show the sunburned area on the upper sides whereas vertical young shoots show discolorations frequently due to sun burning throughout their entire length on the south side. The frosted affected and sunburned trees, produce a succulent twig from the base of the main branches, trunk and from the crown, depending on the injury.

Fruit crops can be protected from frost and freeze damage by using active and passive methods for frost protective with help of frost forecasting. The proper method of frost/freeze protection must be chosen by farmer for the particular site. For successful frost/freeze protection, it is necessary to take care and attention in spraying, fertilizing, pruning, and other cultural practices. The problems that are easily handled during the warm day can become difficult and disastrous during a cold, frosty night so farmer must remain active well before the frost season begins. Winter damage is although phenomenafor most trees, if it occurs continuously on tree, will leads to death of the tree immediately, loss of yields and causes a shorter tree life by making the tree more susceptible to pests and disease. So it is necessary to follow mitigation strategies to reduce the loss.

References

Attaway, J. A., 1997. A history of Florida citrus freezes. Florida Science Source. *Inc. Lake Alfred, FL. ISBN 0-944961-03-7.*

Bagdonas, A, Georg, J. C., and Gerber, J. F., 1978. *Techniques of frost prediction and methods of frost and cold protection.* WMO.

Buszard, D. 1981. Winter injury to apple trees. *Mac donald J.*, 42(11): 9–11.

Czyncsyk, A., 1979. Effect of M.9, B.9 and M.26 rootstocks on growth, fruiting, and frost resistance of apple trees. *Fruit Sci. Rep.***6**:143–152.

Embree, C. 1988. Apple rootstock cold hardiness evaluation.*Compact Fruit Tree***21**:99–105.
    Gagnon, B., Desjardin, Y., and Bédard, R. 1990. Fruiting as a factor in accumulation of carbohydrates and nitrogen and in fall cold hardening of day-neutral strawberry roots. *Journal of the American Society for Horticultural Science,* **115**(4), 520-525.

Krewer, G., 1988. Commodity information - Small Fruit pp. 2-13, *in:* J.D. Gibson (ed). *Cold Weather and Horticultural Crops in Georgia; Effects and Protective Measures.* Extension Horticulture Department, University of Georgia, Publication No. 286.

Martinson, T. E., and White, G. B, 2004. Estimate of crop and wine losses due to winter injury in the finger lakes.

Phillips, D. 1990. Atlantic farming weather: Is it more or less variable today than in the past? Pages 64-85 in R. J. Gordon, ed. Workshop on the application of climate and weather information to the farm. Nova Scotia Agricultural College, Truro, NS.

Powell, A. A. and Himelrick, D. G., 2000. Methods of freeze protection for fruit crops. *Alabama Cooperative Extension System*, 1-9.

ProebstingJr, E. L. and Mills, H. H. 1978.Low temperature resistance [frost hardiness] of developing flower buds of six deciduous fruit species. *Journal-American Society for Horticultural Science (USA).*

Quamme, H. and R. Brownlee. 1997. Cold hardiness evaluation of apple rootstocks. *Acta Hort.***451**:187–193.

Quamme, H., A. Cannon, D. Neilsen, J. Caprio, and W. Taylor. 2010. The potential impact of climate change on the occurrence of winter freeze events in six fruit crops grown in the Okanagan Valley. *Can. J. Plant Sci.* **90**:85–93.

Wang and Wallace, 2003. Susceptibility of fresh fruits and vegetables to freezing injury. In: K.C.

Gross, C.Y. Wang and M. Saltveit(eds). The commercial storage of fruits and vegetables, and Florist Nursery Stocks.USDA Handbook Number, No. 66.

Wildung, D. K., Weiser, C. J. and Pellett, H. M, 1973. Cold hardiness of Malling clonal apple rootstocks under different conditions of winter soil cover. *Can. J. Plant Sci,* **53**(2), 323-329.

Whiteman, T. M., 1957. Freezing points of fruits, vegetables and florist stocks. *USDA Market Research Report,* No. 196. 32p.

Yao, S., Luby, J. J., and Wildung, D. K, 2009. Strawberry cultivar injury after two contrasting Minnesota winters. *Hort Technology,* **19**(4): 803-808.

3

Bearing Habit of Fruit Crops

Madhubala Thakre and Nayan Deepak G

Division of Fruits and Horticultural Technology, ICAR-IARI, New Delhi

In nature, plantspecies flowers produce fruits and seeds and the primary objective is to reproduce them. Plant species those are unable to produce viable seeds, reproduce themselves by other means. With a common of these processes they vary in various aspects of flowering and fruiting. The variations may be due to their different growing habits, requirement of different environments, differences in their basic nature etc. Here we are going to discuss different types of bearing habits do exists in fruit crops.

Growth Habit

Growth is quantitative irreversible increase in the existing size of say plant or a shoot or simply a leaf etc. Fruit crops can be classified on the basis of their growth habits in two major classes. One is evergreen and second is deciduous. Evergreens include those plants which bear their leaves throughout the year. Examples are mango, litchi, guava and orange. Whereas, deciduous trees shed their leaves during a part of year (when weather is adverse *i.e.* during winter). Mostly temperate fruits belong to this category. The presence or absence of leaves makes very big difference in their physiology, their bearing season and pattern etc. These two categories also make a basic group to study them separately with respect to many aspects *viz.* source and sink relationships, pruning etc.

Bearing Habit

The bearing habit of a species can be described by the location and types of buds which produce flower and fruit (Gardner *et al.,* 1952). Now the question is that why do plant bear at a particular location? As plants attains proper nutritive condition particularly accumulation of certain carbohydrates upto desired level, any bud can differentiate to flower bud and set fruit. In various plant species, different buds located at different locations (on a shoot) varies to attain this condition. The justification of formation of fruiting buds on particular locations in a particular species is that at those locations the nutritive and other conditions are more favourable for flower bud formation. Since all the buds can be considered as potential flower buds. So, flowers, inflorescence and finally fruits will bear wherever buds are borne.

They can be located

1. Terminally on long or short growths

2. Laterally in the axils of the current or past season leaves

3. Adventitiously from any point on the exposed bark of limbs, trunks or roots (Gardner *et al.*, 1952).

If we not consider the location of the fruit bud, whether it is present terminal or lateral. After it unfolds it may give rise three basic type of flower-bearing structures:

(1) It may contain flower parts only and develop a single flower (peach) or a flower cluster (cherry) without leaves.

(2) It may be a mixed bud and develop a short or long leafy shoot terminating in an inflorescence (apple).

(3) It may be mixed and develop a short or long leafy shoot bearing flowers or flower clusters in some of its leaf axils (persimmon) (Gardner *et al.*, 1952).

When we combine the above mentioned bud locations and bud type, there will be six groups mainly. Fruits may be classified under these six groups according to their bearing habits.

Classification of Fruit Plants According to Bearing Habits

Six distinct bearing habits, the classification being based upon the location of the fruit buds and the type of flower-bearing structure to which they give rise (Gardner *et al.*, 1952).

Group- I: **Fruit buds borne terminally, containing flower parts only and giving rise to inflorescences without leaves** (Examples: Mango and Loquat)

Bearing habit of this group can understand very clearly form mango. In mango, after harvesting the shoots grows and matures and produces panicle terminally. Growth of lateral branches situated below panicle is continued. In case of accident to terminal panicle, then some of axillary buds may differentiate flower parts.This bearing habit is not found with any of the common deciduous fruits.

	Fruit Buds Terminal	Fruit Buds Lateral
Flower bud containing flower parts only	**Group-I:** Fruit buds borne terminally. It contains flower parts only and produces inflorescence without leaves.(Fig.1) Examples: Mango, Loquat	**Group-IV:** In this group, fruit buds produce laterally. They contain flower parts only and giving rise to inflorescence without leaves or if leaves are present their size is reduced. (Fig.4) **Examples:** Peach, Plum, Apricot, cherry, Almond, Walnut (staminate flower), Pecan (staminate flower), Date, Coconut, Citrus fruit, Plumcot, Current, Gooseberry
Flower bud mixed, Flowering shoot with terminal inflorescence	**Group-II:** Fruit buds borne terminally, unfolding to produce leafy shoots that terminate in flower clusters. This bearing habit is characteristics of most of the pome fruits.(Fig.2) Examples: Apple(principally) Pear (principally), Quince Walnut(P.F.), Pecan (P.F.).	**Group-V:** Fruit buds borne laterally, unfolding to produce leafy shoots that terminate in flower clusters (Fig. 5). Examples: Litchi, Black berry, Rasp berry, Dew berry, Grape, Filbert, Blue berry, Cranberry (European), Cashewnut, Brazilnut, Pond apple (and various other annonaceous fruits), Apple (occasionally), Pear (occasionally).
Flower bud mixed, Flowering shoot with lateral inflorescence	**Group-III:** Fruit buds borne terminally, after unfolding they produce leafy shoots with flowers or flower clusters in leaf axils. This might be called an incomplete terminal bearing habit for the fruit itself is not borne terminally, but is lateral to the growths upon which it appears. However, the flower buds are terminal. The terminal buds of the flowering shoots may differentiate flower parts for the following year's production or new buds may develop from lateral leaf buds (Fig.3). Examples: Pomegranate, guava, tropical almond, olive, *Eugina*.	**Group-VI:** Fruit buds borne laterally(or psudoterminally), unfolding to produce leafy shoots with flower clusters in the leaf axils (Fig. 6). Examples: *Jujube* (Ber), Persimmon, Mulberry, Fig, Cranberry (American), Chestnut, Pistachio nut, Starapple, Avocado, Olive (Partly).

Group- II: **Fruit buds borne terminally, unfolding to produce leafy shoots that terminate in flower clusters**

[Examples: Apple (principally), Pear (principally), Quince, Walnut (P.F.), Pecan (P.F.)]

This type of bearing habit is characteristic of most of the pome fruits. It can be understand very well with example of apple and pear. Majorly, apple and pear, bears on spurs (commercial crop of good quality and quantity). These spurs bear fruit buds terminally. These spurs bear for many years (varies according to varieties). New spurs originate from lateral buds on the shoots of the preceding season and very occasionally from latent of adventitious buds on the trunk or older limb. However, in some varieties of apple which is young and vigorous, many of long shoots form terminal flower buds. But, they contribute very less to total commercial crop of that tree.

Group-III: **Fruit buds borne terminally, unfolding to produce leafy shoots with flowers or flower clusters in the leaf axils**

[Examples: Guava, pomegranate, tropical almond, olive, *Eugina*].

This is an incomplete terminal bearing habit because the fruit itself is not borne terminally, but is present laterally on the growth. However, the flower buds are terminal. The terminal buds of the flowering shoots may differentiate flower parts for the following year s production or new buds may develop from lateral leaf buds. In guava, the fruit buds are formed on short shoots (current season growth) and flower and fruits produced in the leaf axils.

Group-IV: **Fruit buds borne laterally, containing flower parts only and giving rise to inflorescences without leaves or if leaves are present they are much reduced in size.** [Examples: Peach, Plum, Apricot, cherry, Almond, Walnut (staminate flower), Pecan (staminate flower), Date, Coconut, Citrus fruit, Plumcot, Current, Gooseberry].

This type of bearing habit can easily understand by examples of peach and citrus fruit (Kinnow mandarin). In Kinnow mandarin, the fruit buds borne laterally on old season shoot and after emergence it produces flower bud (s) and few leaves. In case of peaches, the shoot of previous season produces flowers on laterally without leaves.

Group-V: **Fruit buds borne laterally, unfolding to produce leafy shoots that terminate in flower clusters.**

[Examples: Litchi, Black berry, Rasp berry, Dew berry, Grape, Filbert, Blue berry, Cranberry

(European), Cashewnut, Brazilnut, Pond apple (and various other annonaceous fruits), Apple (occasionally), Pear (occasionally)].

Bearing habit of group –V can understand by example of litchi. Litchi produces fruit buds laterally which produces leafy shoots which ultimately produces panicles terminally. In case of grapes, the fruit buds borne laterally which produces shoot and this shoot actually bears berry cluster terminally (which makes fit for this group) but as shoot grows further, its growth pushes fruit cluster in one side and it gives appearance as it is situated laterally and opposite a leaf.

Group –VI: **Fruit buds borne laterally (or pseudoterminally), unfolding to produce leafy shoots with flower cluster in the leaf axils.**

[Examples: *Jujube* (Ber), Persimmon, Mulberry, Fig, Cranberry (American), Chestnut, Pistachio nut, Star- apple, Avocado, Olive (Partly)].

Ber is very good example to understand this type of bearing habit. Ber produces various current season shoots from old pruned shoots laterally. These branches bear flower clusters in the leaf axils. After fruit ripening, the leaves and fruits fall and ultimately the branch falls. Buds for the next season crop will produced by strictly vegetative branches. It indicates dimorphism in branching of ber. First type of branches are permanent in mature and forms basic framework for a tree. Second types of branches are deciduous in nature and they dry out after producing flowers and fruits in a season.

Apart from these six main groups, there are three more classes, having distinctness to classify in separate groups.

Group VII: In this type of bearing habit, the fruit buds can be terminal and lateral, inflorescence is generally terminal. The fruit crops discussed in group II and IV can cite here. But it is more convenient to discuss them separately in group II and IV.

Group VIII: This group has very different kind of bearing habit in which fruit buds are adventitious and directly on the trunks, main and smaller limbs and even on the exposed leaves. Examples are cocoa, jackfruit, jaboticaba, cambuca.

Group IX: This type of bearing habit found in fruits which have more or less herbaceous growth habit. In these types of fruit plants fruit buds are present in the axils of the leaves. This group includes the passion fruit (*Passiflora*), the papaya (*Carica papaya*) and many others with a more or less herbaceous type of growth.

Classification of Bearing Habit Considering only the Position of Fruits

Fruits can be classified on the basis of position of fruits on a shoot i.e. whether they are situated terminally, axillary and cauliflorous.

Terminal bearing: Mango, litchi, pineapple, banana, loquat etc. **Arillary bearing:** Guava, ber, apple, papaya, orange, coconut etc. **Cauliflorous bearing:** Cocoa, jaboticaba, jackfruit.

Classification of bearing habit on the basis of shoot maturity on which fruit buds are going to produced:

Another classification of bearing habit is based on the bearing on shoot maturity i.e. old season shoots or current season shoots.

Bearing on old season shoots: Mango, Litchi, Apple, pear, peach, plum etc.

Bearing on current season shoots: Guava, orange, papaya etc.

Relation of Growth Habits to Position of Fruit Buds

There is very close relation between growth habit and fruiting habits. It can be understand by these examples. Fruit plants with terminal fruit buds have a restricted growth habit. In each flowering season, after flowering, the fruits present on the terminal portion. It results in the growth of laterals and restriction of further terminal growth. Finally it gives more compactness to tree. Plants in which fruit buds are borne either terminally (apple) or laterally (sweet cherry) on short growths or spurs are generally more compact as compared to those (like the peach or grape) in which fruit buds are borne on long shoots.In fruits, in which the fruit buds are borne laterally, the resulted extent of compactness depends upon the position of lateral shoot. Whether, it forms on basal, medium or distal portion.

Training and Pruning

Training

When a fruit plant is grown for commercial purpose, there is need to be design growth of that plant in such a manner which results in maximum utilization of all resources (light, space) with sustained production. This is achieved through training. It is practiced during the initial period of establishment of a fruit tree and it decides the form of tree throughout the life of the plant. To train a plant pruning (removal of parts of a tree) is practiced or in other words, training is pruning to control form. Here, formnot only refers to shape of the plant but it also includes the number, orientation, relative size, and angle of branches.The training starts after establishment of a plant in the field. There are basically three systems of training viz. central leader system, open centre system and modified leader system.

1. *Central leader system:* In this system of training, the main stem is allowed to grow. As a result, it is higher than other branches. Mechanically,

it provides much strength to plant. But, it makes difficult different horticultural operations (harvesting) and also affects quality of fruits by shading of upper branches to lower branches (Fig.7).

2. **Open centre system:** In this system of training, the development of vertically growing shoots are stopped by giving thinning out cut to all the branches growing beyond or below a point. So that, it forms „vase shape , it makes available maximum sunlight to all branches. It needs to be pruned severely and constant efforts to maintain it. Moreover, such trees are structurally weak and branches are more likely to break (Fig.8).

3. **Modified leader system:** This is combination of above two types of system. In this system, main stem is allowed to grow but upto a certain height (similar to other branches). For this, the main stem is headed back. Well spaced branches with wider crotches are maintained and weak, intermingling and branches with small crotch angle removed. It makes tree strong as well as with reachable height (Fig.9).

One important point in training is the height of branching. In fruit trees, the main stem kept branch free upto a certain height. This height depends upon the planting distance. When planting distance (for a particular fruit crop) is recommended one then the standard branching height will be 75-90 cm. But, as density increases, the height of branching also decreases and the extent of decrease depends upon the species. One more important point is that, if insufficient branches are available to cover a particular area of plant then branching can be encouraged by pruning. While removing the unwanted branches, the crotch angle is an important consideration. Preferably, branches with wider crotch angles should be selected, as they are able to bear heavy loads of fruits and high wind velocity. Branches with narrow crotch angles are weak because of the lack of continuous cambium and the enclosure of bark and the formation of wood parenchyma in the crotch. Apart from the above mentioned training systems, some other types of training systems are available specific to fruit crops like in case of grape bower system, kniffin system and in case of peach *tatura* trellis system etc.

Pruning

Pruning refers to removal of parts of a tree, especially shoots, roots, limbs, buds or nipping away of the terminal parts. It is done to make a tree more productive and bear quality fruits, increase longevity of the tree, make it into more manageable shape and to get maximum returns from the orchard (Chadha, 2002). Pruning will not only restore balance between shoot and root system, but will also maintain growth and vigour of shoots by allowing only fewer growing points to grow vigorously and regulate the crop (Dubey *et al.*, 2001).

Pruning is done with two broad objectives: first, to regulate the shape of a tree and second, to enhance the production and quality of fruits. Within these broad objectives, pruning is employed to achieve any one of the following or the other similar purposes. These are to: restore shoot: root ratio, prevent formation of weak crotches, regulate the number and location of main scaffold limbs, remove crossing and interfering branches, remove water sprouts or sucker, regulate growth, vigour and direction of shoots, open up the tree canopy, regulate growth and fruiting, induce regular bearing, control size of plants for high density planting, regulate exposure to sunlight, increase the size of fruits facilitate orchard operation, increase the size of fruits, facilitate orchard operation, increase the age of trees, rejuvenate old and senile orchards, control diseases and pests and obtain specific forms (Singh, 2010). Pruning can be performed to remove selected limbs, portion of limbs or large areas of the canopy. The most common implication of pruning is to control trees size, synchronization of vegetative and reproductive cycles but, definitely it can be used to stimulate off-season flowering and fruit production.

There are basic two types of pruning techniques one is heading back and the second is thinning out.

1. ***Heading back:*** It consists of cutting back the terminal portion of a branch to a bud (Fig. 10). Heading back destroys apical dominance and promotes growth of lateral branches. The extent of formation of laterals depends upon the species and the distance from the tip to the cut. It encourages spreading habit of growth and produces a bushy and a compact plant. The cut is made above that bud which is desired to grow. Suppose we want to fill the inner space in a particular part of a tree then the cut should be made above the bud whose direction of growth is inner side of tree.

2. ***Thinning out:*** It is complete removal of a branch to a lateral or main trunk (Fig. 11). Thinning out encourages the longer growth of shoot left after removed branch. Thinning out results in an open tree structure and it provides a larger size as compared to compactness.

Pruning with reference to major fruits: When we talk about the pruning of a particular fruit crop, the bearing habit forms basis for this. Pruning practices for major fruit crops are as follows:

1. **Mango:** Mango bears on old shoots and produces panicles terminally. Hence, pruning is performed after harvesting as early as possible, just to remove unwanted branches and malformed panicles if any. So that, the sprouted shoots will get enough time to sprout.

2. **Orange:** Orange bears on new shoots arise from old shoots during spring season flush. Pruning performed after harvesting as early as possible to remove unwanted and dried branches. It is not recommended to prune an orange tree drastically at a time.

3. **Guava:** Guava bears on current season shoots in the axils of leaves. In guava pruning in performed for two objectives. First one is to pruning to correct tree shape and second one is for crop regulation. When pruning is performed to correct tree shape, it should be performed after fruit harvesting, to remove unwanted branches. For meadow orchards of guava pruning performed three times i.e. in January- February, May-June and September. When pruning is performed for crop regulation it should in the last week of April to first week of May, when the flowering for rainy season crop is more than 50%. The pruning method varies as per growers objective.

4. **Grapes:** Under north Indian conditions, in grapes pruning practices from mid-December to mid-January. It bears on current season shoots from old one. So, particular number of buds kept to facilitate growth of new shoots. Number of buds varies as per variety. Beauty Seedless requires 2-3 buds per cane; Perlett, Delight require 3-4 buds per cane; Pusa Urvashi, Pusa Navrang, Pusa Aditi, Pusa Trisar require 4-6 buds per cane and Pusa Seedless, Kishmish Charni, Thomson Seedless require 9-12 buds per cane.

5. **Ber:** Ber bears on current season shoots. In ber, one single node can produce several flowering branches. These branches bear solitary flowers in the leaf axils. As fruit ripe then after harvesting or dropping on the ground the whole branch dries up and fall on the ground. In the next year the the flowering will occur on the vegetative branches which produces after pruning only. Hence, pruning is routine operation kin ber to get profitable production. So, there is dimorphism in ber. The productive branches are deciduous in nature and the other one who forms the permanent network of the tree.

6. **Apple:** Apple has different type of varieties with respect to their bearing habit. In India, maximum spur bearing varieties are commercially grown. Generally spur remains productive upto 4-5 years. Suppose it is four years for a variety then 25% of the total number of spurs should be pruned. So that, after 4 years grower will get sufficient spurs foe production.

7. **Peach:** Peach bears on the current season shoot of last year. Suppose, one shoot of peach is pruned in December. After breaking of dormancy,

it will produce vegetative shoot (bear on basal part as this shoot is of previous season) which will bear in the next summer. The tip and basal portion of peach shoot is of vegetative in nature. So, the tip should be pruned to produce shoot for next year bearing and the middle portion of that shoot will produce fruits in the same season.

Pruning is based on the bearing habit fruit tree. Hence, for a pruner it is must to have proper knowledge of bearing habit.

Fig. 1: Mango panicles (Bearing habit: Group-I)   **Fig. 2:** Apple flowers (Bearing habit: Group-II)   **Fig. 3:** Guava flowers (Bearing habit: Group-III)

Fig. 4: Kinnow sprouting and flowering (Bearing habit: Group-IV)

Fig. 5: Litchi flowering (Bearing habit: Group-V) **Fig. 6:** Ber flowering (Group-VI)

Fig. 6: Baser flowering (Group-VI)

Fig. 7: Central leader system

Fig. 8: Open centre system

Fig. 9: Modified leader system

Fig. 10: Heading back

Fig. 11: Thinning out

References

Chadha, K. L. 2002. Hand book of Horticulture. Indian Council of Agricultural Research. 41.

Dubey, A. K., Singh, D. B. and Dubey, N. 2001. Deblossoming of summer season flowering of guava (*Psidiumguajava* L.) by shoot pruning.*Progressive Horticulture,* **33**(2): 165-168.

Gardner, V. R, Bradford, F. C. and H. D. Hooker, 1952. The fundamentals of fruit production. McGraw-Hill Book Company, INC., New York, Toronto, London. 536-554.

4

Growth and Development of Fruit Crops

Nayan Deepak G

Division of Fruits and Horticultural Technology, ICAR-IARI, New Delhi

Growth is the fundamental property of all living organisms. It is an irreversible permanent increase in size, volume or mass of a cell or organ or whole organism accompanied by an increase in dry weight. Where, development is sum total of growth and differentiation. It is governed by both environmental and internal factors. The development of a plant is highly complex phenomena (Table 1).

Table 1: Difference between growth and development

Growth	Development
It is addition in length/thickness.	It includes differentiation of organs including reproductive organs.
Growth may occur without development.	Development may occur in absence of growth.
It is more or less quantitative process.	It is a qualitative process.

Type of Growth

Growth can be classified into different types based on different aspects. Growth is of two types:

(a) Indefinite/ Unlimited/ Indeterminate growth – Growth exhibited by root, stem and branches.

(b) Definite/ Limited/ Determinate growth – Growth exhibited by leaves, flowers, fruits etc.

Based on Number of Reproductive Phases in Life, Classified Into

(a) Monocarpic/ Determinate species: Monocarpic plants flower only once in their life. Only one reproductive phase - Annuals

(b) Polycarpic/ Indeterminate species: Polycarpic plants flower every year in particular season. Species have more than one reproductive phase in life. Most of the fruit crops are polycarpic in nature.

Phases of Growth

Growth comprises of three phases Cell division, cell enlargement and cell maturation/differentiation. Total growth period is divided into three stages initial lag phase- slow; middle lag phase / exponent phase- rapid; final – steady phase- gradually decline Growth at all phases is not equal. Growth is completely stops at the end. The total time taken by a cell or organ to complete all phase is called "grand period of growth". Graph plotted of time and organ growth is called "grand period of growth curve" (Fig. 1).

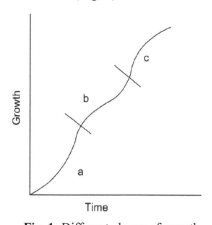

Fig. 1: Different phases of growth
1: Lag/ Initial Phase, 2: Lag/ Exponential Phase, 3: Final /Steady Phase

Growth can be measured by auxanometer in terms of length. While growth rate is measured in terms of absolute growth rate (AGR) and relative growth rate (RGR).

(i)   **AGR** – Total growth of each plant or organ per unit tree

   AGR = dw/ dt

   Where, dw = increase in weight and dt = change in time

(ii) **RGR** – Growth per unit time expressed per unit of weight or volume

   RGR = dw/ dt x 1 / Wo

   Where, dw = difference between final and initial weight; Wo = initial weight and t = time in days

Type of Growth Curves in Fruit Crops

1. **Sigmoid/ Single sigmoid curve** – Fruit undergoes slow enlargement at the early (a) and last stages of growth (c), while growth is considerably faster during the middle development stage (b). Ex: Apple, pear, pineapple, banana, avocado, almond, strawberry, loquat, date palm, papaya, mango and lemon

2. **Double sigmoid curve** – Three stages are seen

a. Ovary, nucellus and integuments of the seed grow rapidly, but the embryo and endosperm grow little.

b. Embryo and endosperm grow rapidly, but the ovary does not increase much in size, sclerification of the pit also begins and embryo achieve full size by the end and the amount of endosperm material increases greatly.

c. A new surge of ovary growth begins and continues to fruit ripening

Ex: Peach, plum, apricot, ber, raspberries, fig, blackberry, blueberry, cherry, pecanut, persimmon, guava, grapes, olives, etc.

3. Triple sigmoid curve – Five stages are seen

a. Initial rapid growth, seeds reaching full size (0-9 weeks)

b. Slow growth, seeds hardens and start to colour, first very large respiratory response to ethylene (9-12 weeks)

c. Rapid growth, seeds become dark brown, response to ethylene increases (12-17 weeks)

d. Very little growth, seeds dark brown, softening starts, soluble solids starts to increases, respiratory response to ethylene rises to a maximum and then decreases (17-21 weeks)

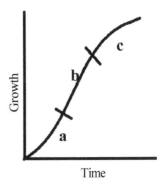
Fig. 2: Single sigmoid curve

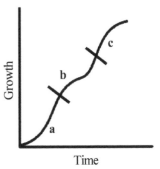
Fig. 3: Double sigmoid curve

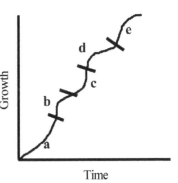
Fig. 4: Triple sigmoid curve

e. A smaller but significant growth increase to approximately final size. The fruit matures, the seeds becoming very dark brown and free in the tissues. Ex: Kiwi fruit.

Every plant has two distinct phases or development namely vegetative and reproductive. In the vegetative phase, i.e.leaves, roots, shoots etc., and in reproductive phase flowers and fruits are produced. In annuals, these phases occur for a short period, each phase occurring only once and that too within the same season. In the biennials the vegetative phase occurs for one season and in the following season the reproductive phase. In such cases, the two phases are intercepted by a period of rest in winter in the northern latitudes.

The vegetative growth is for several years depending upon the variety, planting material, species, growing conditions etc. This is known as the pre-bearing period. After pre-bearing plant enter to reproductive phase starts bearing flowers and fruits. Once the perennial tree complete its pre-bearing period, the cycle of vegetative and reproductive phases might occur in each season or the reproductive phase may occur hand in hand with vegetative phase. In some trees the vegetative and reproductive growth may occur in alternate years known as alternate bearing. In some other trees, the vegetative growth may occur for two, three or more years followed by reproductive growth for one year and again vegetative growth for two or more years. This is known as irregular bearing.

If vegetative shoot is arise from a bud is known as a vegetative bud, while the buds which produces floral parts only or a shoot with flowers is known as the flower bud/fruit bud/blossom bud. In case, the bud produces only floral parts, it is known as a pure flower bud and if it produces a shoot with flowers and leaves it is known as mixed bud.

Classification of Fruit Plants on the Basis of Flowering Habit

According to Kozlowski (1971) there are four groups of fruit plants based on their season of flowering.

a) **Year round/ Ever flowering species:** Plants produce flowers throughout the year irrespective of photo thermal changes. Examples are Fig and Papaya.

b) **Seasonal flowering species**: Plants produce flowers in the particular season, any variation in environmental factor during favourable season leads to continuance of vegetative phase. Examples are Guava, Litchi, Apple and Pear.

c) **Non Seasonal flowering species:** Plants produce variation of flowers from plant to plant and from part to part. The variation is more prominent around the equator and the plant becomes seasonal, as they grown away from the equator. Examples are Mango, Cashewnut and Coconut

d) Gregarious flowering species

Flowering occurs at particular times in a year is influenced by the atmospheric causes like rainfall in drier period or chilling induces indefinite flowering. Differentiation, formation may takes place at congenial time but flower opening and anthesis require an impulse due to temperature. Example: Quince

The physiological and morphological changes that occur in a vegetative bud in its preparation to change over to the reproductive phase or to become a flower bud may be called as the flower bud initiation. Further, development changes that occur in an initiated flower bud leading to the formation of the embryonic flower inside the bud are known as flower bud differentiation. Both the steps of initiation and differentiation together may be called as flower bud formation. Depending upon species or variety or kind of fruit tree, there may be varying periods of interval between the formation of the flower bud and the actual production of flowers.

Phases of Fruit Development

Fruit development can generally be considered to occur in four phases:

a)   Fruit set,

b)   a period of rapid cell division, c) a cell expansion phase, and

d)   ripening/ maturation.

Seeds are ripened ovules; fruits are the ripened ovaries or carpels that contain the seeds. Fruits develop from organs of the flower and thus involve differentiation or redifferentiation of preexisting organs. Evolutionarily, floral organs represent modified leaves and so the fruit is also a modified leaf.

True and False Fruits

True fruits	False fruits
Fruit derived from a single ovary. Outside of the fruit is called the pericarp and develops from the ovary wall. During fertilisation an embryo is formed in the ovule. This results from the fusion of male and female reproductive cells (a nucleus in the pollen grain and a nucleus in the female egg cell in the ovule). There are other nuclei in the pollen grain and the egg cell and these also fuse and form a structure known as the endosperm. This becomes a food store for the developing seed.	Fruits are composed of tissues derived from flower parts (e.g. sepals, .petals, stamens, stigma and style) other than the ovary or from more than one ovary. Ex. Apple andStrawberry

Fruit Ripening

Ripening represents the shift from the protective function to dispersal function of the fruit. Ripening occurs synchronously with seed and embryo maturation, as described in the lecture on embryo development. Ripening involves the softening, increased juiciness and sweetness, and color changes of the fruit. Fleshy fruits are either climacteric or non-climacteric. Ethylene is a major regulator of the ripening process.

Growth and development of some fruit crops

Banana	Inflorescence emerges 9-12 months after planting and fruits are ready to harvest after 3-4 months
Ber	Flowering during Sep-Nov in North India and August in Western India. Flowering period is for 50-70 days. Fruits are harvested 120-18- days after blooming. Fruits are harvested during Nov-Apr.
Guava	Vegetative propagated and seedling ones starts commercial yield after 2-3 years and 4-5 years respectively. Flowering and fruiting twice in North India, while thrice in South India and Western India.Bahar treatment was followed in guava for flower regulation. There are three bahars *i.e.* Ambe bahar, Mrig bahar and Haste bahar.a. Ambe bahar – Plants were stressed during Dec-Jan which leads to flowering and fruiting during Feb-Mar and Jul-Aug respectively.b. Mrig bahar – Plants were stressed during Apr which leads to flowering and fruiting during Jun-Jul and Nov-Dec respectively.c. Haste bahar – Plants were stressed during Aug-Sep which leads to flowering and fruiting during Oct-Nov and Feb-Apr respectively.Fruits require 5 months from full bloom to maturity
Litchi	Vegetative propagated plants take 3-5 years in North India and 6 years in SouthIndia for flowering. Seedlings take 10-15 years.N. India – Flowering starts last week of Jan- first week of Feb and fruit ripens during May-Jun.S. India – Flowering starts in Dec and Fruit ripens in Apr-May. Fruits take 105-120 days for maturity from flowering.
Mango	Commercial yield is obtained after four-five years old in vegetative propagated trees, whereas seedlings take seven years. Flower bud differentiation occurs during October – February. Flowering period is for 14 – 21 days. Availability of fruits is from February to July. Fruits are ready for harvesting in 5 – 6 months from flowering.
Papaya	Flowering and fruiting throughout the year. Starts flowering five months after planting and harvesting 9-10 months after planting.
Sapota	Economic yield obtained after 7 years in vegetative plants and 10 years in seedlings. Sapota flowers throughout the year in tropical condition with two main seasons are July-Nov and Feb-Mar. Require 7-8 months for harvesting from flowering.

References

Childers N. F, 1983. Modern fruit science, Orchard and small fruit culture, Freeman USA. Edward Francis Dumer, 2013. Principles of Horticultural physiology. CABI Science, pp 39-40. Eng Chong Pua and Michael R. Davey, 2009.Plant development biology – Biotechnological perspectives, Volume 1. Springler, pp 306-307.

Gardener V. R, Bradford F. C and Hooker H. P, 1952.Fundamentals of fruit production.

Harlan K. Pratt and Michael S. Reid, 1974. Chinese Gooseberry: Seasonal patterns in Fruit Growth and Maturation, Ripening, respiration and the role of ethylene. *J. Sci. Fd Agric*, 25, 747:757.

Kozlowski T. V, 1971. Growth and development of trees. Vol II. Academic press, New York: 313-319.

Rajput C. B. S and Pattanayak B. K, 1985.Training and pruning of tropical and subtropical fruits.Naya prakash, Calcutta, 56-57.

Stephen G Pallady, 2010. Physiology of woody plants. Academic Press, Science, pp 94-9.

5

Method of Irrigation in Fruit Crops

Amit Kumar Goswami

Division of Fruits and Horticultural Technology, ICAR-IARI, New Delhi

In the words of *Leonardo the vinci*, "water is the driver of life". It is said to be the liquid of life or elixir of life. Water availability is essential to consider as commercial orchards must have reliability of supply and cannot produce optimum yields without irrigation. Water management is one of the largest and most important inputs into an orchard. The single environmental component which determines the type of fruit plants to be selected for a specific ecological situation is the moisture status of the region. Different fruit plants preferring diverse moisture status have wide peculiarity in their anatomical and morphological structures as well as various modifications in their water absorbing organs. They are also seen to have varied capabilities to acquire and utilize water efficiently. In the plant system, the rate of dry matter production has a definite relation with the rate of transpirational loss of water and this transpiration ration is calculated on the basis of unit of water required to produce unit of drymatter. However, while calculating the water requirement of fruit plants, a number of factors, in addition to water use efficiency, are taken in to consideration. Naturally water requirement of a particular fruit plants varies with the agro-ecological condition of different places. Moreover the same cannot be equal in an established orchard and a newly planted one. In established orchard, the trees are grown in community where improved water use efficiency lower down the water requirement of the trees.

The growth and productivity of a fruit plants as well as profitability of the orchard enterprise depend on the moisture relations and irrigation practices. Irrigation is very important in fruit crops as sufficient moisture must be maintained in the soil for obtaining the optimum yield of good quality fruits. For a profitable orchard enterprise, a well planned irrigation system and efficient water management practice having utmost importance. The aim of irrigating a fruit tree should be to wet the entire root zone without allowing any wastage of water beyond the root zone. The irrigation systems have to be properly devised so the water requirement of the trees is met at the minimum expenditure without any wastage of water. By definition, irrigation is the artificial application of water to the land or soil. It is used to assist in the growing of agricultural crops, maintenance of landscapes, and revegetation of disturbed soils in dry areas and during periods of inadequate rainfall. And efficient water management refers to artificial application of water *i.e.* irrigation in crop root zones in case of soil moisture deficit and removal of water *i.e.* drainage from the root zone in case of excess so as to provide the crops a most optimum soil moisture regime for best production. It means how best we are in irrigating and dewatering our fields so that plants are protected from stress as well as water logging. Various factors such as soil type, crop type, planting density, water quality, irrigation equipment and economic factors such as the capital and operating costs will all determine the ultimate decision for choosing the type of orchard irrigation systems.

The choice of the irrigation method depends on the following factors

(i)   Size, shape, and slope of the field.

(ii)   Soil characteristics.

(iii)   Nature and availability of the irrigation water supply.

(iv)   Types of crops being grown and age of the trees.

(v)   Initial development costs and availability of funds.

(vi)   Preferences and past experience of the farmer.

Thus the system of irrigation must be decided in relation to varying orchard condition. There are different methods of irrigation in fruit crops and every method has some advantages and disadvantages.

Methods of Irrigation

Irrigation water can be applied to crop lands using one of the following irrigation methods

(i)   Traditional /Surface irrigation

   a)   Uncontrolled (wild or free) flooding method

   b)   Border strip method

   c)   Check basins method

   d)   Modified/ring basin

   e)   Furrow method

   f)   Pitcher Irrigation

(ii)   Advanced / Pressurized Methods

    a)   Sprinkler irrigation

    b)   Trickle (Drip) irrigation

    c)   Talca Irrigation Management System (TIMAS)

    d)   Partial Root Zone Drying (PRD)

    e)   Bubbler irrigation

Surface Irrigation

In all the surface methods of irrigation, water is either ponded on the soil or allowed to flow continuously over the soil surface for the duration of irrigation. Although surface irrigation is the oldest and most common method of irrigation, it does not result in high levels of performance. This is mainly because of uncertain infiltration rates which are affected by year-to- year changes in the cropping pattern, cultivation practices, climatic factors, and many other factors. As a result, correct estimation of irrigation efficiency of surface irrigation is difficult. Application efficiencies for surface methods may range from about 40 to 80 per cent.

(a)   **Uncontrolled flooding:** This system is very simplest and easiest to practice. The water is allowed for irrigation without making any beds, basins or any other structure. In „uncontrolled , wild or „free flooding, water is applied/ flooded to the orchards without any preparation of land and without any levees to guide or restrict the flow of water on the field. The advantage of this method is the low initial cost of land preparation. This method is 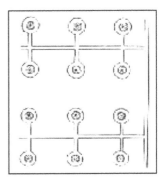 suitable when water is available in large quantities, the land surface is irregular, and the crop being grown is unaffected because of excess water. In this method, water is brought to field ditches and then admitted at one end of the field thus letting it flood the entire field without any control. Uncontrolled flooding generally results in excess irrigation at the inlet region of the field and insufficient irrigation at the outlet end. Application efficiency is reduced because of either deep percolation (in case of longer duration of flooding) or flowing away of water (in case of shorter flooding duration) from the field. The application efficiency would also depend on the depth of flooding, the rate of intake of water into the soil, the size of the stream and topography of the field. This method is not economical as well as not suitable for the crops which are very sensitive to water logging like papaya.

(b) **Border Strip Method**: This is a controlled surface flooding method of applying irrigation water in the orchard. In this method, the farm is divided into a number of strips. These strips are separated by small ridge or raised area.

Water from the supply ditch is diverted to these strips along which it flows slowly towards the downstream end and in the process it wets and

irrigates the soil. When the water supply is stopped, it recedes from the upstream end to the downstream end. The border strip method is suited to soils of moderately low intake rates and low erodibility. This method, however, requires preparation of land involving high initial cost.

(c) **Check-basin Method**: The check basin method of irrigation is based on rapid application of irrigation water to a level or nearly level area completely enclosed by dikes. In this method, the entire field is divided into a number of almost levelled plots surrounded by levees. This method is suitable for a wide range of soils ranging from very permeable to heavy soils. The farmer has very good control over the distribution of water in different areas of his farm. Loss of water through deep percolation (near the supply ditch) and surface runoff can be minimised and adequate irrigation of the entire farm can be achieved. Thus, application efficiency is higher for this method. Besides, there is some loss of cultivable area which is occupied by the levees. Sometimes, levees are made sufficiently wide so that some „row crops can be grown over the levee surface.

(d) **Furrow Method**: Previously furrow system of irrigation is attributed as one of the best systems for irrigating the mature trees. This method is also practiced in newly established orchards. This method is an alternative to flooding. In this method of irrigation, the entire land surface is to construct in small channels along the primary direction of the movement of water and letting the water flow through these channels which are termed „furrows , „creases or „corrugation . Furrows are small channels having a continuous and almost uniform slope in the direction of irrigation. Water infiltrates through the wetted perimeter of the furrows and moves vertically and then laterally to saturate the soil. Furrows are used to irrigate crops planted in rows. Furrows necessitate the wetting of only about half to one-fifth of the field surface. This reduces the evaporation loss considerably. However, the depth, length and width of furrows depend on nature of the soil and spread of root system of the fruit plants. Furrows provide better on-farm water management capabilities for most of the surface irrigation conditions and variable and severe topographical conditions. For example, with the change in supply conditions, number of simultaneously supplied furrows can be easily changed. In this manner, very high irrigation efficiency can be achieved.

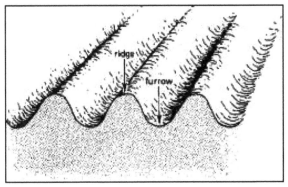

(e) **Ring basin method:** This is very useful method of irrigating the young tree in the orchard. In this method, a circular ring in the periphery is prepared to irrigate the plants. While preparing, care is taken that ring is prepared away from tree trunk towards outer periphery of the tree. In between two ring-basins, a sub channel connecting the ring basin of the tree is prepared. The water flow through central channel and move ahead naturally after flooding two ring basins at a time.

(f) **Pitcher Irrigation**: This system is very suitable for those areas where water scarcity exists. The pitcher filled with water buried in the periphery of individual tree where feeding roots are confined. It is similar to drip irrigation but less expensive to install. The pitchers are the round earthen containers used in rural areas for water storage, ranging from 10 to 20 liters in capacity. This kind of irrigation is

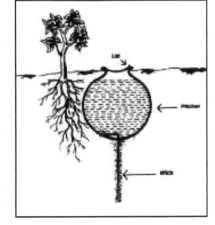

ideal for saplings, promoting deep root growth. Soluble fertilizers can also be mixed with water and applied through the pitcher. If the water used for irrigation has high salinity, the pitcher location should be changed in every 3 years. To increase the depth of irrigation, a wick can be added to the pitcher.

(ii) Pressurized Methods: Advanced Techniques for increasing water use efficiency broadly devided in two groups:

A. Subsurface Irrigation: Subsurface irrigation (or simply sub irrigation) is the practice of applying water to soils directly under the surface. Moisture reaches the plant roots through capillary action. In natural sub irrigation, water is distributed in a series of ditches about 0.6 to 0.9 meter deep and 0.3 meter wide having vertical sides. These ditches are spaced 45 to 90 meters apart. Sometimes, when soil conditions are favourable for the production of cash crops (*i.e.,* high-priced crops) on small areas, a pipe distribution system is placed in the soil well below the surface. This method of applying water is known as artificial sub-irrigation. Soils which permit free lateral movement of water, rapid capillary movement in the root-zone soil, and very slow downward movement of water in the subsoil are very suitable for artificial sub-irrigation. The cost of such methods is very high. However, the water consumption is as low as one-third of the surface irrigation methods. The yield also improves. The conditions which favourable sub irrigation are as follows:

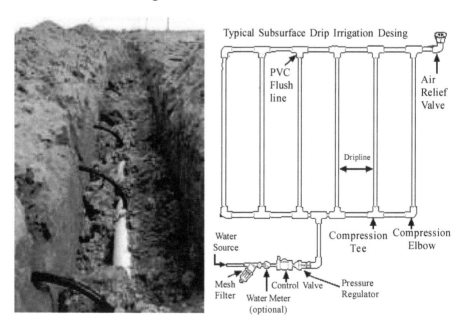

Typical Subsurface Drip Irrigation Desing

(i)   Impervious subsoil at a depth of 2 meters or more,

(ii)   A very permeable subsoil,

(iii) A permeable loam or sandy loam surface soil,

(iv)   Uniform topographic conditions, and

(v)   Moderate ground slopes.

B. Micro- irrigation: A scientific method of irrigation carrying desired water and nutrients direct to the root zone of the plant, drop by drop. Micro-irrigation systems apply water to the tree line of the orchard only, whereas other irrigation methods also wet the traffic-lane between rows. Irrigating only the treeline ensures traffic access to the orchard at all times for such essential operations as spraying and harvesting. The amount of water applied by micro-irrigation systems can be closely managed to match the requirements of the crop. Therefore, the frequency and volume of application are factors that can be used to control the growth of the crop to maximise marketable yield. Fertiliser can also be readily applied through the pipe network of micro- irrigation system so that the nutrient solution is applied directly to the active root zone. This reduces the losses of fertiliser that would occur through percolation or uptake by weeds. The main problem that can occur with micro-forms of irrigation is the blockage of the emitter. Blockages can be prevented provided attention is paid to installing adequate filtration and chlorination systems and operating them in accordance with the manufacturer s instructions. Micro-irrigation has a number of advantages over conventional methods of irrigation. The choice between types of micro-irrigation is not as easy and comes down to choosing with a wide choice of wetting pattern size and shape. There are different systems of micro-irrigation but care has to be taken in selection of irrigation method. Following methods of micro-irrigation are widely used in orchards according the requirement and availability of recourses.

1.   Sprinkler Irrigation

2.   Drip Irrigation

3.   Others system like Talca Irrigation Management System (TIMAS), Partial Root Zone Drying (PRD) and Bubbler irrigation

1. Sprinkler irrigation: Sprinkling is the method of applying water in the form of a spray which is somewhat similar to rain. In this method, water is sprayed into the air and allowed to fall on the soil surface in a uniform pattern at a rate less than the infiltration rate of the soil. This method started in the beginning of this century and was initially limited to nurseries. In the beginning, it was used in humid regions as a supplemental method of irrigation. Sprinkler irrigation usually wets the whole orchard floor. Sprinkler systems offer reasonable control of irrigation run time. Rotating sprinkler-head systems are commonly used for sprinkler irrigation. Each rotating sprinkler head applies water to a given area, size of which is governed by the nozzle size and the water pressure. Alternatively, perforated pipe can be used to deliver water  through very small holes which are drilled at close intervals along a segment of the circumference of a pipe. The trajectories of these jets provide fairly uniform application of water over a strip of cropland along both sides of the pipe. With the availability of flexible PVC pipes, the sprinkler systems can be made portable too. Sprinklers have been used on all types of soils on lands of different topography and slopes, and for many crops.

The following conditions are favourable for sprinkler irrigation:

1. Very previous soils which do not permit good distribution of water by surface methods,

2. Lands which have steep slopes and easily erodible soils,

3. Irrigation channels which are too small to distribute water efficiently by surface irrigation,

4. Lands with shallow soils and undulating lands which prevent proper levelling required for surface methods of irrigation.

There are so many advantages with the sprinkler system of irrigation like low water loss (efficiency up to 80%), saving in fertilizer, suitable for any topography, no soil erosion, better seed germination, free aeration of root zone and uniform application of water. The main difference between sprinkler systems and drip systems of irrigation is the wetting of a larger soil volume by the spray or jet emitters. This occurs by virtue of the water being distributed over a larger area of soil but the drip systems apply water to the one point and rely on the soil properties for distribution of the water. The wetting of a larger surface of soil is important on sandy soils where little lateral movement occurs within the soil and on some clay soils where cracking of the soil is severe. The wetting of a larger soil volume should result in bigger trees but not necessary more productive trees. Wetting a larger soil volume makes for a safer system in case the interval between irrigations is longer and hence, there is less risk from excessive soil dryness. Some disadvantages also exists with this system of irrigation like high initial cost, cannot adopt by ordinary farmers, poor application efficiency in windy weather and high temperature, high evaporation losses, water should be free of debris, equipments need careful handling, physical damage to crops by application of high intensity spray and power requires for running pumping unit etc.

Trickle (Drip) Irrigation

Drip irrigation has enabled farmers, nurserymen and landscapers to conserve water for decades. Drip irrigation, also known as trickle irrigation or micro irrigation or localized irrigation, is an irrigation method that saves water and fertilizer by allowing water to drip slowly to the roots of plants, either onto the soil surface or directly onto the root zone, through a network of valves, pipes, tubing, and emitters. It is done through narrow tubes that deliver water directly to the base of the plant. This is primarily because, in contrast to gravity or sprinkler irrigation, drip irrigation technology applies water slowly and directly to the targeted plant's root zone. In addition, drip irrigation technology has extremely high application uniformity, even when pressures vary from hilly terrain or long lengths of run, or where planted areas are oddly shaped. Trickle irrigation (also known as drip irrigation) system comprises main line, sub mains, laterals, valves (to control the flow), drippers or emitters (to supply water to the plants), pressure gauges, water meters, filters to reduce clogging of the emitters (to remove all debris, sand and clay), pumps, fertilizer tanks, vacuum breakers, and pressure regulators.

The drippers are designed to supply water at the desired rate directly to the soil. Low pressure heads at the emitters are considered adequate as the soil capillary forces causes the emitted water to spread laterally and vertically. Flow is controlled manually or set to automatically either to deliver desired amount of water for a predetermined time or to supply water whenever soil moisture decreases to a predetermined amount. Today it is more important than ever to use water resources wisely and to irrigate intelligently. Consequently, many farmers have turned to drip irrigation and have enjoyed improved profitability by increasing crop yield and quality while at the same time reducing costs from water, energy, labour, chemical inputs and water runoff. Many farmers have also enjoyed significant water and capital investment savings using drip irrigation, while simultaneously improving plant vigour by delivering water and nutrients directly to the plant roots and avoiding unnecessary wetting of plant leaves.

Drip irrigation is the targeted, intelligent application of water, fertilizer, and chemicals that when used properly can provide great benefits such as water use efficiency is maximal so water saving, maintaining high soil-water potential in root zone, partial soil wetting so no interference with agro technical practices, lesser number of weeds, no wetted canopy that may cause disease, even low quality water can be utilized, can be operational under heavy winds, easy application of fertilizer, herbicides and pesticides through drip lines, adaptation to marginal plot, high uniformity in water and fertilizer supply and comparatively pressure requirement is low in this system. In spite of the fact that drip irrigation has so many potential benefits, they re a certain limitation also like sensitivity to clogging, moisture distribution problem and salinity hazards, high cost compared to furrow, high skill is required for design, install and operation and not suitable for closely planted crops.

Others systems like TIMAS uses weather data in combination with soil water content measurements to provide farmers with information required to manage the irrigation scheduling. The Partial root-zone drying (PRD) method of irrigation is a modified form of deficit irrigation, involves irrigating only one part of the root zone in each irrigation event, leaving another part to dry to certain soil water content before rewetting by shifting irrigation to the dry side; therefore, PRD is a novel irrigation strategy since half of the roots is placed in drying soil and the other half is growing in irrigated soil. Originally, the concept of PRD was first used in USA on field cotton in alternate furrow irrigation. Wetting and drying each side of roots are dependent on crops, growing stage, evaporative demands, soil texture and soil water balance.

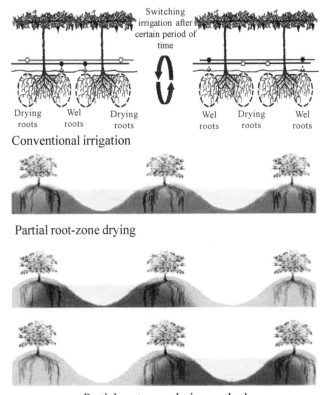

Conventional irrigation

Partial root-zone drying

Partial root-zone drying method

Yet there is little understanding on mechanism of PRD effects on crop growth, therefore, no definite solid procedure exist on determining the optimum timing of irrigation for each side. Wetting and drying each side of roots are dependent on crops, growing stage, evaporative demands, soil texture and soil water balance. Yet there is little understanding on mechanism of PRD effects on crop growth, therefore, no definite solid procedure exist on determining the optimum timing of irrigation for each side. Therefore, in PRD, roots sense the soil drying and induce ABA that reduce leaf expansion and stomatal conductance and simultaneously the roots in wet soil absorb sufficient water to maintain a high water status in shoot. Practically, PRD can be used in different ways depending on the cultivated crops and/or soil conditions, environmental conditions and method of irrigation. PRD has been used by surface and subsurface drip irrigation. Partial root-zone drying irrigation is the novel deficit irrigation strategy that is generally adapted in the last decade to a vast kind of agronomic and horticultural crops to increase the water productivity. However, the amount of saved irrigation water and improved water productivity strongly depends on crop, soil, and site specifications. Moreover, cumulative results revealed that partial root-zone drying irrigation could not be effective in reproductive crops that are sensitive to water stress. Therefore, partial root-zone drying is recommended for irrigation of farms and gardens in arid and semi-arid areas which are suffering from lack of fresh water resources for agricultural production. Partial root-zone drying practices can be viable and advantageous option compared with full irrigation to prevent crop yield reduction when and if there is water shortage or to improve crop quality. It is noteworthy that studies on PRD are still continuing and in future new results will be available from other crop species, probably from horticultural and tree crops with a high irrigation water requirement.

Another less popular system like bubbler irrigation which is a localized, low pressure, solid permanent installation system used in tree groves. Each tree has a round or square basin which is flooded with water during irrigation. The water infiltrates into the soil and wets the root zone. The water is applied through bubblers. These are small emitters placed in the basins which discharge water at flow rates of 100–250 litres/h. Each basin can have one or two bubblers as required. With bubbler irrigation the percentage of the root soil volume wetted is about 80 percent. Bubbler irrigation is mainly applied in fruit tree orchards. Bubbler emitters discharge water on the same spot of ground at high rates. Thus, for a uniform distribution over the basin area, a minimum of land preparation is needed. This Bubbler system having high irrigation application efficiency (up to 75 percent), resulting in considerable water savings, with absolute control of the irrigation water from the source to the tree basin. The entire piping network is buried so there are no field operations problems. The technology is simple and no highly sophisticated equipment is used. This system can be operated by unskilled farmers and labourers.

Quality of Irrigation Water

Surface water, ground water, and suitably treated waste waters are generally used for irrigation purposes. Irrigation water must not have direct or indirect undesirable effects on the health of human beings, animals, and plants. The irrigation water must not damage the soil and not endanger the quality of surface and ground waters with which it comes into contact. The presence of toxic substances in irrigation water may threaten the vegetation besides degrading the suitability of soil for future cultivation.

The various types of impurities, which make the water unfit for irrigation, are classified as:

- Total concentration of soluble salts in water
- Proportion of sodium ions to other ions

- Concentration of potentially toxic elements present in water

- Bacterial contamination

- Sediment concentration in water

References

Bucks, D. A., Nakayama, F. S. and Warrick, A. W., 1982. Principles, practices and potentialities of trickle (drip) irrigation. Advances in irrigation, 1:219-297.

Garg, S.K (2007), Irrigation Engineering and Hydraulic Structures, Khanna Publishers, New Delhi.

Michael, A.M. 1978. Irrigation: Theory and Practices. Vikash Publishing House Pvt. Ltd. New Delhi. Pp. 801.

Nakayama, F.S. and Bucks. D.A. 1990. Water quality in drip/trickle irrigation: A review. irrigation science, 12(4): 187-192.

6

Water and Nutrient Management for Fruit Crops

M. Hasan

Centre for Protected Cultivation Technology, IARI, New Delhi

Water is one of the most important and critical input for agricultural production system. The demand for food grains is increasing day by day due to increasing population. The irrigated agriculture provides the crop water productivity of about 2.5 tonnes per hectare and the overall irrigation efficiency is only about 30%, which is below the world average level. In the above scenario it is necessary to have the irrigation system in which both the crop water productivity and irrigation efficiency increases considerably. Micro irrigation has become the most viable and efficient technology in such a situation. It provides several advantages in the context of crop agronomy, water and energy conservation. Micro irrigation has the potential to achieve the crop water productivity to a desired level of 4 tonnes per hectare and simultaneously maintain the irrigation efficiency above 80%. The total coverage of micro irrigation in the tenth plan is only about 2 million hectare. The task force on micro irrigation (2004) has indicated a potential of 69 million hectare for our country. So there is a tremendous potential available and the coverage of micro irrigation has to be increased to cover more crop and more new areas.

Horticultural sector, which includes vegetables, flowers and fruits, has enormous potential for development owing to varying agro climatic conditions in India. There is tremendous scope for increase in productivity and quality of horticultural crops, which is necessary to compete in this modern competitive world market. There are many production technologies available to increase the per unit vegetable production and also to increase its quality. One of the widely acceptable, reliable and cost effective technologies is pressurized irrigation. Water is one of the most crucial inputs for growing irrigated vegetables in large scale. The increasing population and industrialization has led to decrease in the allocation of water to agricultural sector. Thus water has become a very precious and costly input. Pressurized irrigation includes drip, sprinkler, jet and spray. Drip irrigation is the best available technology for the judicious use of water for growing fruit crops in large scale on sustainable basis (Hasan *et al.*, 2004). Drip irrigation is a low labor intensive and highly efficient system of irrigation, which is also amenable to use in difficult situations and problematic soils, even with poor quality water. Irrigation water savings ranging from 36-79% can be affected by adopting a suitable Drip irrigation system. Drip irrigation or low volume irrigation is designed to supply filtered water directly to the root zone of the plant so as to maintain the soil moisture near to field capacity level for most of the time. Water and fertilizer saving around 25 and 30 percent respectively through drip fertigation system over traditional irrigation system was reported for various fruit crops for Delhi region (Hasan *et al.*, 2006). The Field capacity soil moisture level is found to be ideal for efficient growing of vegetable plants. This is due to the fact that at this level the plant gets ideal mixture of water and air for its development. The device that delivers the water to the plant is called dripper. Water is frequently applied to the soil through emitter placed along a water delivery lateral line placed near the plant row. The principle of drip irrigation is to irrigate the root zone of the plant rather than the soil and getting minimal wetted soil surface. This is the reason for getting very high water application efficiency (90-95%) through drip irrigation. The area between the crop row is not irrigated therefore more area of land can be irrigated with the same amount of water. Thus water saving and production per unit of water is very high in drip irrigation. Sprinkler, jet and spray irrigation simulates rainfall and are useful for specific horticultural crops. Sprinkler irrigation is found to be very useful in orchard for fertigation of mature trees. It is also very efficient in crop like potato and leafy vegetable crops. It has the efficiency in the range of 80-85 %. It is not suitable for the area having high wind velocity and humid condition prevailing for a long duration.

Micro Irrigation System Network has Five Basic Units

- Pumping unit.
- Control head.
- Pipe network (Main and submain pipes).
- Laterals and dripper/sprinkler/sprayer.
- Hydraulic connections.

Water is pumped through the groundwater in most of the cases. Pressurized water is required for the micro irrigation. Normally 5 Hp pump is enough to irrigate the vegetable crops of one hectare and orchard having fruit trees in about 2 hectare. The control head is equipped with primary and secondary filters, pressure gauge, water meter, valves, fertigation tank and pump, hydraulic connections. Gravel filter is required for primary filtration and disc filter is required for secondary filter of groundwater. The minimum pressure required at the head for the working of drip and sprinkler irrigation are 2 and 2.5 bar respectively. Pressure control, non- return, safety and air valves are required. Fertilizer pump, fertilizer tank or venture is required for the supply of nutrients along with water. Main line, sub-main line and lateral pipes are required to supply the water and nutrients simultaneously. Main and sub- main lines are mostly of PVC. The diameter of main line usually varies from 60-90 cm having pressure of about 6 Bar. Sub- main lines are made of either PVC or HDPE having diameter of about 25-60 cm. Lateral pipes are made up of LDPE and its diameter varies from 12-20 mm. In the case of drip irrigation, in- line or on-line drippers are used for final discharge of water to the plants. In line drippers are most commonly used in vegetable and flower cultivation and they are easier to use and install. On line drippers are mostly used in landscaping and orchard crops. The capacity of dripper varies from 2-10 liter per hour and spacing mostly varies from 20-60 cm. Sprinkler head is required at the laterals for discharge of water in sprinkler irrigation. Sprinkler head can be of partially or fully rotating type and its discharge varies from 25-90 liter per hour. Sprinkler system can be of permanent, semi-permanent or shifting type. Sprinkler gun is used to discharge very high quantity of water as required in landscaping, cereal crops and in orchards. Hydraulic connections like Tee, end plugs, elbows, joints, bends etc are required in large quantity in installation and maintenance of micro irrigation system (Fig .1).

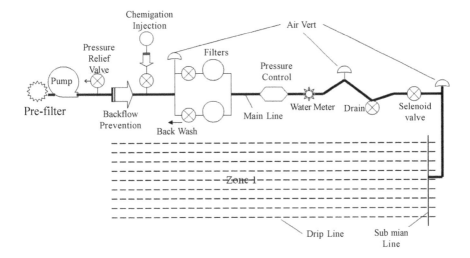

Fig. 1: Layout of micro irrigation system

The usual life of micro irrigation system is said to be 7-10 years. It can be achieved only by regular maintenance of different components of micro irrigation system. The micro irrigation system can become non operational or ineffective if all its components are not maintained regularly. Most of the maintenance jobs can be done very easily without any extra expenditure. Filters, lateral line, main line should be cleaned and flushed regularly.

Types of Pressurized Irrigation Systems

A distinction is made between the two principal micro-irrigation methods, namely, the sprayer or micro-sprinkler, and the drip irrigation system. Sprayers and micro-sprinklers spray the water through the atmosphere and are designed principally to wet a specific volume of soil around individual trees in an orchard. Drip irrigation, on the other hand, represents a paint source of water, and wets a specific volume of soil by direct application of water to the root zone of the plant. The type of drip emitter from the aspect of its discharge and the distribution of the emitters throughout the plot (distances along the drip lateral and between the drip laterals) is dependent on the soil texture and the crop. The drip system is suitable for irrigation of row crops (vegetable and industrial crops) and orchards.

Sprinkler Irrigation

As stated, the object of sprinkler irrigation is to imitate rainfall. Rain spreads water over large areas in uniform manner. The sprinklers, which are mounted generally on aluminum or plastic laterals, spray water by a jet or jets of water (ejected from one or two nozzles) which cause the impact-driven sprinkler to rotate in a circular manner and to spread water over the area according to the radius of through of the jet. The quantity of water accumulated in the immediate vicinity of the sprinkler is generally much greater than that which reaches the soil at the end point of the wetted radius, depending on its particular distribution profile. Hence, in order to obtain high water application uniformity in the field, the sprinklers must be spaced so that the spray from one sprinkler reaches the adjacent sprinkler, i.e., the sprays from adjacent sprinklers must overlap. Spacing of the sprinklers at the correct distances ensures satisfactory uniformity of distribution provided the sprinkler is designed to ensure a regularly shaped distribution pattern and is operated in suitable wind and pressure conditions. Wind adversely affects distribution uniformity. It is therefore recommended to operate sprinkler irrigation systems in windless conditions or in light wind.

The manufacturers specify the optimal operating pressure for each of their sprinkler models. Operating the sprinkler system at the optimal pressure, and given the other above- mentioned conditions, should ensure adequate distribution uniformity. The term application rate, i.e. the amount of water applied to the soil, expressed in mm/hr, is a function of the discharge of the sprinklers and their spacing. The infiltration rate of the soil, also expressed in mm/hr, differs for different soils. In order to ensure proper infiltration of the water into the soil and to prevent surface run-off, it is necessary to design the system so that it gives an application rate, which is less than the infiltration rate of the soil.

Since sprinklers spread the water in a circular manner, it is recommended to use part- circle sprinklers, which enable adjustment of the area wetted by the sprinkler, so as to minimize loss of water in the area bordering the plot and to avoid interference to vehicles which travel on the field tracks which run along the borders of the irrigated fields.

Proper attention to the points listed below in the various stages of field data collection, planning of the irrigation system, and its installation and operation in the field will result in high distribution uniformities and irrigation efficiencies;

(i)   Spacing of the sprinkler s at distances between them, which will enable attainment of high distribution uniformity;

(ii) Irrigation at an application rate less than the infiltration capacity of the soil;

(iii) Operation of the sprinkler system in windless conditions or light winds only;

(iv) Operation of the sprinkler system according to the pressure recommended by the manufacturer;

(v)   Use of part-circle sprinklers in combination with full-circle sprinklers.

(vi) Portable, i.e. hand-move sprinkler irrigation laterals with adequate overlap can be installed to reduce investment costs.

Sprayers and Micro Sprinklers

The micro-sprayer is a device for spreading water through the atmosphere by means of a nozzle and a static spreader platform, whereas the micro-sprinkler spreads the water by means of a rotating, whirling, sprayer device. For this reason, mini-sprinklers with a discharge identical to that of a sprayer wet wider areas. The micro-sprayer/sprinkler is most suitable for below canopy irrigation of orchard trees. The objective is to wet the soil in a limited area around the tree without wetting all the soil area occupied by each tree. Sprayers exist which have a bridge-like device which enables the static spreader nozzle to be replaced by a rotating device, in this manner, it is possible to irrigate the young orchard by means of micro-sprayers and to switch over to micro-sprinklers when the orchard matures, requiring wetting of a larger soil volume. Micro-sprayer and micro-sprinkler systems can attain maximum irrigation efficiency of 85%.

Basic Components of Drip Irrigation System

A typical system layout consists of the following basic components.

Control head: It includes pump, filters, fertilizer applicator, water meter, pressure/flow regulating valves and controller for automation.

Filters: These are essential components required to remove suspended materials from the pumped water that might clog the drippers. Different types of filters are used in the system. Gravel and graded-sand filters are cylindrical tanks that are filled with fine gravel and sand of selected sizes. These are mainly used for filtering out sand and organic materials after the pumping. Screen and disk filters are the least expensive and most efficient means for filtering water. These are generally used as the second stage filters.

Fertilizer applicators: These are used to inject fertilizer, systemic insecticides/algaecides, acids and other liquid materials into the water being supplied through drip system. They are of 3 types, namely, fertilizer tank, venturi system and fertilizer pumps.

(i)   Fertilizer tank: A metallic tank is provided at the head of the drip irrigation system for applying fertilizers in solution along with the irrigation water. The tank is connected to the main irrigation line by means of a by pass line. Some of the irrigation water is diverted from main irrigation line in to the tank. This bypass flow is created by the pressure difference between the entry and exit points of the tank. This application method is simple in construction and operation with low requirement of electricity. However, the application of fertilizer during the fertigation schedule is not constant. Therefore, it does not permit a precise control over the fertilizer concentration.

(ii) Venturi system: It consists of a built-in converging section, throat and diverging section. A suction effect is created at the converging section due to high velocity, which allows the entry of the liquid fertilizer in to the system. This system is simple in operation and a fairly uniform fertilizer concentration can be maintained in the irrigation water.

(iii) Fertilizer pump: This type of pump is operated either with electricity or with water. It draws fertilizer solution from a tank and pumps it under pressure in to the irrigation system. It provides a precision control on the fertilizer application. However, it is expensive and needs skilled operation.

Pressure/flow regulators: These are control valves that are actuated either manually or electro- hydraulically to regulate flow and pressure in the drip system.

Controllers: These automatic – mostly micro-processor based – devices are used to provide stop/start signals to pump and valves/regulators. The actuating signal may either be time or volume based. In more advanced technological modes, these gadgets are controlled by soil moisture sensors placed in the plant root zone.

Pipe lines: The water conveyance from the control head to the emitter / dripper is generally categorized in three units as follows:

Main pipe: This is the main carrier of water from the source (after the control head). It is further connected to sub-mains or manifold. These are usually rigid PVC and HDPE pipes with pressure rating of about 10 kg/cm^2.

Sub-mains: This is the portion of pipe network between the main pipe line and the laterals. These are also PVC / HDPE / LDPE pipes with pressure rating of 6-7 kg/cm^2. Diameters of main and sub-main pipes are selected based on the water requirements of the farm area.

Laterals: These flexible pipes of HDPE or LLDPE ranging from 10 mm to 20 mm diameters are the ones which are spread over the field in a specified layout. Designed to carry water at about 3 kg/cm^2, these pipes are provided with point-source emitters or drippers spaced along it.

Emitter or dripper: This is a device designed to dissipate the hydraulic pressure and to discharge a small uniform flow of water, drop by drop, at the given place. Different types of emitters have been developed based on the pressure dissipation and flow mechanisms. Emitters, generally made of poly-propelene material, are either mounted on the lateral (on-line type) or fixed inside the lateral (in line type). The emitters are available in different discharge rate capacities ranging from 1 to 8 litres/ hr. Selection of emitter of given discharge rate depends on the type of crop and soil.

Fertigation for Fruit Crops

Fertigation is the process in which fertilizers can be applied through the system with the irrigation water directly to the region where most of the plant roots develop. It is done with the aid of special fertilizer apparatus (injectors) installed at the head control unit of the system, before the filter. The element most commonly applied is nitrogen. However, application of phosphorous, potassium and other micro-nutrients are common for different horticultural crops. Fertigation is a necessity in drip irrigation. The main objectives of fertigation are;

1. Uniform and timely application of fertilizers.

2. Water and nutrient saving.

3. Optimizing yield.

4. Quality improvement.

5. Minimizing pollution

The rational for fertigation are as under:

- Irrigation and fertilizers are the most important management factors through which farmers control plant development and yield.

- Water and fertilizers have important synergism which is very well used in fertigation.

- Timely application of water and fertilizers can be controlled through fertigation.

Principles of Fertigation

It is to feed the plant in appropriate time, quantity and location. These three things can be controlled through fertigation. The plant yield and the quality depend on all these three factors.

Advantages of Fertigation

- Amount and concentration of nutrient can be adjusted according to the stage of development and climatic considerations.

- Deeper penetration of nutrients into the soil.

- Avoiding ammonia volatilizing from soil surface

- Application restricted to the wetted area where the active roots are concentrated.

- Reduced time fluctuation in nutrient concentrations.

- Crop foliage is kept dry, thus retarding the development of plant pathogens and avoiding leaf burn.

- Allows fertilization in the rainy season when the soil is in wet condition without stepping on it and destroying the structure.

- Convenient use of fertilizers.

- Remote control operation.

- Convenience in saving manpower.

- Low losses in transportation and storage.

- The system may be used for additional applications.

Fertigation Scheduling for Horticultural Crops

Nutrient management in fertigation is important for increasing the crop productivity and also quality of produce. Plant needs nutrients throughout their growth stages. For the newly planted fruit trees the dosage for first year fertigation is 10% of the recommended dose of fully mature fruit trees and it will gradually increase by 20% of the dosage for the succeeding years. In horticultural crops, these stages vary from crop to crop viz initial and final stage, vegetative stage, flowering stage, flowering and fruiting stage etc.(Table 1 and 2).

Table 1: Approximate cost of drip irrigation system under different plant spacing for fruit crops

Plant spacing, m x m	Approx Cost of drip system, Rs/ha
12 x 12	50,000
10 x 10	60,000
8 x 8	70,000
6 x 6	80,000
4 x 4	90,000
2 x 2	1,00,000
1 x 1	1,25,000

Water Requirement of Fruit Crops

It depends on climatic, soil and plant based parameters. It is calculated from pan evaporation based method by multiplying pan evaporation and crop coefficient. It is relatively simpler method and is commonly used for determination of water requirement of fruit crops. It can also be calculated by FAO based Penman-Monteith method, which is a very precise method having many complex parameters. The soil physical properties based method uses Field capacity, wilting point and bulk density for crop water estimation. It also uses root zone depth for total crop water estimation.

Calculation of Water Requirement

1. Pan Evaporation based method

WR = A x B x C x D x E

Where:

WR = Water requirement (lpd /plant)

Table 2: Water saving and yields enhancement due to drip irrigation

Sl.No.	Crop	Yield (q/ha)		Irrigation (cm)		WUE (q/ha/cm)		Advantage of drip irrigation (%)	
		Surface	Drip	Surface	Drip	Surface	Drip	Saving	Increase in yield
1.	Banana	575.00	875.00	176.00	97.00	3.27	9.00	45.00	52.20
2.	Grapes	264.00	325.00	53.00	28.00	5.00	11.60	47.00	23.10
3.	Papaya	130.00	230.00	228.00	73.00	0.60	3.20	67.90	76.90
4.	Pomegranate	34.00	67.00	21.00	16.00	1.62	4.20	23.80	97.00
5.	Watermelon	82.10	504.00	72.00	25.00	5.90	20.20	65.30	513.9

A = Open Pan evaporation (mm/day)

B =Pan Factor (0.7); this may differ area wise

C = Spacing of plant (mm^2)

D = Crop factor (factor depends on plant growth, value range 0.4-1.0)

E = Wetted Area (0.3-0.7 for widely spaced crops)

The total water requirement of the farm plot would be WR x No. of Plants.

The irrigation water requirement is determined using average season wise pan evaporation data for the area. The average figure of 20 or 30 years is taken into account while doing the calculations.

2. Penman Method

Penman in 1948 proposed a method for calculating the evaporation from a water surface taking into considerations several climatic parameters. It is taken that the evaporation rate is a function of the resistance to the movement of vapour through the profile of air above the evaporating surface, and depends on the vapour pressure gradient and the local wind speed.

$E_0 = (\Delta(R_n-G)/L + \gamma E_x)/ (\Delta+\gamma)$

Where,

E_0= evaporation from a free water surface (mm/day)

R_n = net radiation $(cal/cm^2/day)$

E_x = isothermal evaporation (mm/day)

L = latent heat of evaporation

Δ = slope of the saturation vapour pressure temperature curve

γ = psychrometric constant

This equation requires more climatological data and is also more accurate than the above two methods. Considering the mechanics of movement of water vapour through leaf stomata, Monteith (1992) modified this equation and it becomes Penman-Monteith equation. It is the best available method for calculating ET. It is valid for the dry crop completely shading the ground.

$ET = [\Delta(R_n - G)/L + cp\rho a\ ea\text{-}ed/Lra]/ (D+\gamma*)$

Where,

ET = evapotranspiration rate from and dry crop surface.

cp = specific heat of dry air at constant pressure

ρ_a = density of moist air

e_a-e_d = vapour pressure deficit

r_a = aerodynamic resistance of the crop

Intake and utilization of water: It depends on irrigation scheduling and water budget. Irrigation scheduling consists of two important components namely crop water requirement and irrigation frequency. The intake and utilization pattern of water for different fruit crops depend on above parameters.

The concept of water budget is used to know the water balance and intake vs utilization of water in the orchard. The water budget equation correlates irrigation requirement in the orchard with crop evapotranspiration, effective rainfall and irrigation system losses.

Irrigation requirement = ET_C- Effective Rainfall + Irrigation system losses

Crop evapotranspiration (ET_C) is the sum of evaporation from soil and transpiration from leaves.

Effective rainfall is the part of total rainfall stored in the soil for effectively utilized by plants.

Response of fruit plants to varying conditions of soil moisture: The fruit crops response is sensitive to different regime of soil moisture. Field capacity soil moisture regime is known as the optimum soil moisture condition for plant growth and better yield. It can be maintained in the soil for prolong duration with the help of drip irrigation. Optimum soil moisture is required to be maintained during the critical stages of different fruits for getting high yield and better quality. Fruit crops are adversely affected if soil moisture is maintained around the wilting point. The soil moisture in the fruits can be maintained around field capacity level with the help of tensiometers. Deficit irrigation in the form of Regulated deficit irrigation (RDI) and Partial root zone irrigation (PRD) is the latest concept of soil moisture based irrigation for orchard in water scarcity areas.

Pathological conditions associated with excess and deficiency of moisture: Pathological condition in the orchard is also correlated with excess and deficiency of moisture. Wilting is found to be the most common symptom due to moisture deficiency. Fungal related disease is common in the orchard having excess and uncontrolled moisture. Optimum moisture condition is one of the prerequisite for the healthy orchard.

Major Recommendations for Drip Irrigation and Fertigation in Orchard

1. Drip Irrigation and Fertigation is found to be useful for fruit crops in water scarcity areas for enhancing quality and yield.

2. Mulching in combination with Drip Fertigation gives good result in fruit crops.

3. Location specific indigenous design of pressurized drip Fertigation is required for farmers.

4. Indigenous fertigation scheduling for fruit crops in simpler terms is required for

5. Training and Demonstration is necessary to promote drip Fertigation for farmers.

References

Hasan, M. and N. P. S., Sirohi. 2006. Irrigation and fertigation scheduling for peach and citrus crops. *Journal of Agricultural Engineering*, 43(4): 43-46.

Hasan, M., Singh, A. K. and N. P. S. Sirohi. 2004. Performance and evaluation of different irrigation scheduling methods for peach through efficient fertigation system network. *Acta Hort* 662: 193-197.

7

Endogenous and Exogenous Factors Responsible for Flowering in Fruit Crops

K. Usha, M.K. Verma and Nayan Deepak G.

Division of Fruits and Horticultural Technology, ICAR-IARI, New Delhi-110 012

Introduction

Fruits play a vital role in improving human health which are called as protective foods rich in vitamins, minerals, fats, oils, proteins and improves in economy of producers, exporters and industrialists. Fruit crops can be classified into two broad categories *i.e.* ever green and deciduous trees.

a) Evergreen fruit trees grow under tropical and subtropical climate and they are not dependent on chilling requirement for floral bud differentiation and flowering. Ex: Mango, Citrus, Guava, Sapota, Papaya etc.,

b) Deciduous fruit trees require dormancy during winter season and chilling units for floral initiation and have specialized modifications to protect sensitive growing tissue from unfavourable conditions. Ex: Apple, Pear, Peach, Plum, Apricot, Cherry, Grapevine, Walnut, Almonds etc.,

The number and quality of flower buds leads to flowering and fruit yield, which depends on control of numerous external and internal signals. Major pathways to flowering include environmental induction through photoperiod, vernalization, gibberellins, autonomous floral initiation and aging responding to endogenous cues that determine the flowering time. The conditions controlling floral transition are determined by certain complex growth correlations and difficult to know about the certainty of branch that will produce flowering buds. There is need to understand the salient features and differences between the phenologies of deciduous and evergreen fruit trees and impacts of climate change on the flowering. This chapter helps in understanding the exogenous and endogenous factors that control flowering in fruit crops.

Factors Effecting Flowering in Fruit Trees

i. Influence of Juvenility

There will be a series of qualitative transitions in fruit trees during their life-cycle in response to environmental and endogenous cues and require a long juvenile phase before producing flowers. Dramatic transition changes occurs in fruit tree from the vegetative to reproductive stage involves the transmission of the integrated signal of floral induction to the floral meristem identity genes and floral morphogenesis which are governed by a complex genetic network. Both genetic and environmental factors are known to influence the duration of the juvenile phase which can vary from 5–12 years. A number of biochemical changes like differences in peroxidase, esterase isozymes protein phosphorylation and endogenous hormone levels of auxin, abscisic acid, cytokinin and ethylene are involved in the juvenile-to-adult phase transition. Gibberellic acid (GA) plays promotional and repressive effects depending on tree species. Juvenile shoots capable of producing vegetative and mature shoots produce flower buds in a sequential manner. Floral initiation in some tree crops and largely regulated by maturity of the most recent flush and characteristics of the shoot was affected by the timing of vegetative growth. In litchi the vegetative growth is during recurrent flushing, with the interval between successive flushes dependent on the existing weather conditions.

ii. Age of wood

Depending on the specific shoot age and position of buds in shoot there are variations in flowering patterns in species and their cultivars. The cyclic initiation of shoots on dormant stems, whether vegetative or reproductive, is common to many fruit species. The timing of flush development for flowering is important because bud release for vegetative or reproductive growth can only occur from mature flush. Buds are receptive for only a small portion of the flush development cycle and floral induction requires inductive temperatures to coincide with bud release. The induction switch is governed by the interaction of a putative floral promoter and an age-regulated vegetative promoter possibly gibberellins in leaves or buds at the time of initiation. Flowering of mango occurs in tropical areas that lack cool night temperatures only when shoots become sufficiently aged.

iii. Tree architecture on flowering

Reproductive strategies are connected to axis polymorphism that determines plant architecture. Axis polymorphism results from meristem expression, activity and is related to axis orientation or axis length. Differentiation of axis size affects stem (length, diameter and dry mass), leaves (area and dry mass) or both and expresses the adaptive behaviour of fruit tree. Dimorphism in dioecious fruit species is due to different reproductive behave such as males or as females. Differences in the secondary sexual characteristics such as leaf area, inter node size, form of canopy and in primary and secondary growth characteristics in male and female plants of same species. The axis length, the predominance of stem components over the foliar components, referred to as axialization, determines the vegetative status of the axes in apple, cherry and some tropical species. The higher and lower axialization leads to vegetative and floriferous axils respectively. Axialization is low during the first stages of growth causes the rapid autotrophy of the plant, whereas it is high at the adult stage when the plant explores the environment in bramble fruit, whereas in cherry, the axialization progressively decreases as the tree ages, in parallel with the reduction of axis length and the development of flowering on the axes. When branching processes differ between dimorphic axes, the tree architecture and form are affected by the number and the position of each kind of axis. Apical growth units that have larger stem and leaf area are more likely to flower, fruit and branch. In apple, the quadratic relationship between growth unit diameter and flowering was dependent on the position and factors other than carbohydrates regulated flowering.

iv. Position of buds on flowering

There will be varying in the types of shoots that produce flowers and where on these shoots the flowers are borne in fruit crops. Flowering and vegetative growth interact was influenced by the location of the flowers and timing of floral initiation. Litchi tends to produce inflorescences from terminal buds, whereas avocado produces inflorescences from both terminal and axillary buds. In walnut three different types of flowering habits *i.e.,* terminal, intermediate and lateral. The main bearing unit was spurs in apple, pear, apricot, cherry and almond. Similar to apple, pear and sweet cherry initiate flowers on specialized spur structures as well as on current season s growth, whereas peach and apricot initiate flowers in lateral buds of the current season s growth. Apical flowering requires the end of shoot growth before panicle initiation can take place in several tropical and subtropical fruit crops.

v. Dormancy on flowering

Bud dormancy in perennials is a complex process that enables plants to survive long periods of adverse conditions including the extremes of drought, cold, and heat. Individual stems of fruit trees are dormant for most of the time and growth occurs as periodic, ephemeral flushes of shoots emerging from apical or lateral resting buds before returning to a quiescent state. Usually a period of quiescence in the canopy caused by environmental conditions was observed for flower induction in subtropical and tropical evergreen species. Generally, reproductive flushes occur after extended periods of stem rest in the low-latitude tropics or immediately following periods of cool night temperatures in the higher latitude of tropics and subtropics. Declining photoperiods and temperatures reduce shoot extension growth and cause initiation of apical buds to protect the apical meristem during late summer. Particular environmental or endogenous signal perceived within the bud, induces and maintains these buds in a state of endodormancy. In grapevine floral initiation occurs in uncommitted primordia of developing latent buds, destined for dormancy and subsequent release in the spring. The uncommitted primordia produce tendrils as well as inflorescences, but tendrils are produced when growth is initiated without undergoing winter dormancy. The chilling and forcing temperatures during the endodormancy and ecodormancy are perceived by the floral primordial and the consequent chilling and heat requirements are main factors in the breaking of both dormancy stages. Uneven bloom is due to insufficient chilling. The release of pre-existing floral or vegetative buds in the spring after winter dormancy is dependent on the chilling requirements in temperate deciduous species. Interesting character in avocado that is panicle initiation prior to bud dormancy indicating that rest is not a prerequisite for the transition to flowering. In mango for panicle initiation and bud growth, exposure to cool inductive temperatures is necessary. Warm winter locations with insufficient winter chilling leads to poor bud break in kiwifruit.

vi. Water stress on flowering

Drought and water stress increase the intensity and synchrony of flowering not only in subtropical areas where low temperature triggers the flowering process, but also for some cultivars in tropical areas. This response is due to a lesser competition with exuberant vegetative growth. Moderately low water potentials delay shoot initiation through reduced turgor thus extending the age of stems. Despite the role of water stress, a period of cool and dry weather for flowering and long stress period reduce flower size and fruit set. Citrus and pomegranate producers following water withholding strategy for out of season flowering.

vii. Temperature on flowering

Temperature is one of the important factors influencing flowering at various stages of flower initiation to development of various parts. Cool temperature induce flowering in several tropical and subtropical fruit trees while in temperate deciduous temperature can affect the intensity of floral initiation. Flowering can be induced by exposure to low temperature in citrus, litchi, macadamia, avocado, orange and Olive. High temperature increases inflorescence production in grapevine, while temperature of 21°C and below increases tendril production. High root temperatures can inhibit floral initiation in litchi implicating by the roots and long-distance signalling or heat transfer *via* the transpirational stream.

In mango and litchi the number and ratio of hermaphrodite-to-male flowers changes in relation to temperature and other environmental conditions. In avocado, temperatures of 10/7°C (day/night) were effective in inducing flower buds, whereas at 25/20°C (day/night) regime was not. In the subtropical trees litchi, avocado, and macadamia, flowering is dependent on bud release during cool florally inductive temperatures. As hours of low temperature inductive conditions increase, the type of new growth arising from axillary buds changes from vegetative to mixed (flowers and leaves), to generative (leafless with at least one flower). Six different shoot types, i.e., totally vegetative, totally flowering, mixed panicle, two transition stages (vegetative-to flower and flower-to-vegetative), and chimeral (flowers on one side and leaves on the other) may develop depending mainly on temperatures and their duration in mango.

viii. Photoperiod on flowering

Photoperiodic induction is a common mechanism in herbaceous species but rarely observed in perennial fruit trees. Southern high bush blueberry flowers in response to short days of 8 h, but not under long days or short days with a 1 h night interruption. Both the time to flowering and floral initiation were decreased by short days of 9 h compared with 15 h, but these effects may be related to differences in photosynthetic period and daily carbon assimilation rather than to photoperiod in avocado. In kiwifruit buds enter dormancy in response to shortening day length. Exposed shoots produced more floral buds and inflorescences per shoot than shaded shoots. Mango, banana and plantain are day neutral for floral induction. Photoperiods of less than 12 h are associated with a slowing the rate of bunch initiation that is independent of temperature expressed as growing degree days and contributes to seasonal variations in banana flowering. Apical dominance and competition for light, contribute to variation in flowering in ratoon crops, due to marked effect of plant density. Strawberry floral induction occurred at the end of summer or the beginning of fall when the days are short and temperature is low.

ix. Hormone levels on flowering

Critical balance of auxin, cytokinin and gibberellin compounds may regulate shoot initiation in plants. Polar auxin transport from a dominant sink was suggested as a possible mobile signal affecting flowering. The application of auxin, polar transport inhibitors resulted in flowering induction. Endogenous cytokinin levels in buds increase at the onset of floral initiation and differentiation. GA inhibit floral initiation by reducing the rate of node development, reduced bud appendage production, bud size and both delayed and reduced the transition of buds from vegetative to floral differentiation. CK conjugates stored in wood and bark, might serve as a source to build up this high CKs/ (auxin+GAs) ratio which is essential for successful flower induction. The parallel reduction in ABA and IAA levels in the bud would cross-talk between the ABA and IAA signaling pathways.

x. Alternate bearing

Fruit trees exhibit two major reproductive strategies, i.e., first phase, plant produces sufficient amount of vegetative growth to support production of an ample number of flowers during the subsequent year. Fruit crops like fig, some orange and grapefruit cultivars are regular bearers, characterized by a relatively stable multi annual yield and usually possess efficient mechanism(s) to control excess fruit production. In second type, trees bear a heavy fruit load (ON-Crop) in one year, which inhibits return bloom and vegetative growth the next year that leads to low yield (OFF-Crop) and high vegetative growth. Alternating bearing trees are usually characterized by low self-thinning ability examples are mango, apple, olive, pistachio, mandarins and so on. Early observations linked carbohydrate and nitrogen metabolism to the process. A remarkable increase in the expression of genes encoding proteins associated with calcium-dependent auxin polar transport and a reduction in bud endogenous auxin levels following de- fruiting and its role in alternate bearing. Higher levels of IAA in ON buds reflect their inability to distribute IAA efficiently via the Ca^{2+}-dependent PIN-based polar auxin transport mechanism. In addition, efficient auxin removal from the bud appears to be a key component in transforming the ON bud into an OFF bud.

xi. Fruit load on flowering

Fruit load might act at different developmental stages such as flowering induction, transition of the shoot apical meristem and subsequent stages of flower development and bud break. Higher the fruit load causes lower number of nodes sprouted in spring. The developing fruit provides a strong sink for photo assimilates and was therefore thought that depletion of photo assimilates, especially carbohydrates from the bud prevents flowering induction. The inhibitory effect of fruit in flower formation depends on the number of fruits developed and the harvest date. Heavy fruit load prevents recognition of the low-temperature flowering inductive signal and/or blocks later stages of inflorescence, such as bud break. As fruit load and flowering intensity is not a linear function, there might be a threshold value for crop load and above which flowering is strongly inhibited.

xii. Carbon: Nitrogen on flowering

Floral initiation is strongly influenced by carbohydrate availability. Flower-bud formation occurs when the carbohydrate supply predominates and shoot growth is stimulated when nitrogen supply is high. Depletion of carbohydrate and/or nutrient resources influences the development of floral buds for the next year. Heavy bearing in fruit species can cause depletion of stored carbohydrates in shoots. Variations in carbohydrate availability among spurs suggest that branches and spurs function as semi-autonomous organs during most of the year except for early spring. Branches are dependent on the tree for water and mineral nutrient supply but carbohydrate sinks located on an individual branch are largely supplied by the sources located on the same branch, making each branch a semi-autonomous system. The proteins of photosynthetic machinery represent the majority of leaf nitrogen which is directly related to photosynthetic capacity in C/N theory.

xiii. Climate change on flowering

Climate change is a significant and lasting change in the distribution of weather patterns over periods ranging from decades to millions of years. The global atmospheric temperature is increasing @0.3 to 0.6% annually and in next 100 years global temperature may increase by 1.1 to 6.4°C with an average of 1.8-4.0°C. Climate change induces changes in temperature, moisture, humidity and incidence of pest and diseases in fruit crops. In most fruit crops higher temperature decreased the days interval required for flowering. Temperature not only influences the development of various parts of flowers but also determines the type of inflorescence. In citrus, more leafless floral shoots are produced at cooler temperature (20/ 15°C day/night) and higher soil and air temperature enhanced production of leafy floral shoots. Rain, heavy dew or foggy weather during the blooming season stimulate tree growth but interfere with flower production and encourage diseases of the inflorescence. Anthesis was affected due to extreme temperatures and drought. The lack of chilling resulted in abnormal patterns of bud-break and development of trees with delayed flowering date and an extended flowering Period. High winter temperatures, which are 11 or 12°C above seasonal norms and high winter air temperatures causes too early flowering in pistachio trees. Climate changes are visible clearly in the shifting of apple cultivation from lower elevations to higher altitudes in Himachal Pradesh. Overall decrease of about 2–3% in yield. In a warming environment, frost risk decreases as temperature less frequently fall below 0 °C. Climate warming sufficient to induce phenological advances should similarly shorten the frost season. If the last frost dates advance at a rate equal to or faster than the phenological shift, then leaf buds, flowers and resultant fruit yields will experience a decreased frost risk. If a drought or intense heat/cold damages trees and their buds, tree care, pruning and waiting for favourable weather next year is the only option.

Different Management Strategies for Flowering

a) Managing tree architecture for flowering in fruit crops

For flowering transition in many tropical and subtropical fruit crops require reduced vegetative activity. Alterations include water stress, drought, branch manipulation, mechanical shoot and root pruning, application of hormones, chemical growth regulators, manipulating crop load, fertilizer management and regulated deficit irrigation. Winter pruning increases vegetative growth and decreases flowering. The effects of pruning also vary with the architectural level, the complete removal of 1-year-old short lateral shoots tends to stimulate growth of the remaining shoots and fruit set of adjacent inflorescences. The pruning strategies should be based on the removal of flowering shoots at a young stage of growth, with less pruning of old branches which increases leaf area of remaining shoots and light interception by the tree canopy through a decreased shoot density and a better distribution of shoots in space. By following this method improve the physiological autonomy of the flowering shoot, especially during the period preceding June drop, by enhancing carbon assimilation necessary for current fruit and bourse- shoot growth and flower initiation in the terminal bud of the shoot. Shoot growth and flower-bud formation is obviously affected by shoot orientation, *i.e.* placing shoots in a horizontal position increases flower-bud formation and reduces growth. This knowledge provides insights for a better understanding of the determinants of flowering and fruiting.

b) Breaking dormancy

A range of rest-breaking chemicals like dinitro-ocresol (DNOC), thiourea, gibberellic acid (GA3), various surfactants, various oils, potassium nitrate, hydrogen cyanamide and combinations of chemicals were using. Dormex (H2HCN), Volk oil and potassium nitrate are few other examples of commercial dormancy breaking chemicals. Hydrogen cyanamide is highly effective in breaking the dormancy of grape buds and kiwi vines. H2O2 content was increased and came at peak in late dormancy, when endodormancy was broken. Thiourea (1.5%) was most effective to break bud in apple, plum, peach and apricot. Thiourea (2%) induces flowering in „Pathernakh pear in India. In apple, exogenous application of the NH4 ion, rather than the NO3 ion, promotes flowering. Reduced vegetative growth and low GA content induced flowering in avocado in response to low temperature. A period of vegetative dormancy needed to initiate floral buds can be induced by low temperature, water stress, withholding fertilizers, cincturing and auxin sprays in litchi.

c) Overcoming alternate bearing in fruit crops

The alternate bearing tendency of individual spurs, whole trees, or orchards; necessitates the de-synchronization of the on/off cycle of fruit trees. Fruit load control through chemical or manual thinning has received particular attention in many perennial crops. In apple 30 days after full bloom, chemical thinning is commonly practiced because fruit to fruit competition and the detrimental effect of fruit on floral initiation are still low. Thinning agents such as the benzyladenine (BA) or the Naphtalene acetic acid (NAA) may present a threat for the environment and restricted which leads to the demand for alternative strategies among which the selection of new cultivars with self thinning properties or planting regular bearing varieties.

d) Managing hormone levels in the tree

Reduced levels of endogenous GA have been correlated with floral initiation in citrus, litchi and GA biosynthesis inhibitors have improved flowering in mango, longan, litchi and macadamia. Triazoles and other classes of plant growth retardants, which inhibit gibberellins biosynthesis, promote strong and out-of-season flowering in younger trees. Paclobutrazol is substitute provided by mild water stress or low nitrogen to obtain flowering on younger stems. There is an increase both endogenous cytokinin levels and floral initiation in,,Japanese pear after application of the growth retardant maleic hydrazide. Cytokinin application as zeatin increases flowering on spurs and replaced the need for leaves in floral initiation on defoliated spurs. Daminozide increases flower bud formation in apple and has decreased GA tissue concentrations and increase cytokinin tissue content. Winter GA sprays that can be used to regulate flowering in citrus and minimize the effect of biennial bearing. Use of IBA can change the bud ratio in twigs and regulate the flower numbers in staminate and pistillate inflorescence.

e) Nutrient management

With proper balance between Carbon and nitrogen promotes flowering in fruit trees. Stress on large crops, improper fertilization and no fertilization after fruit harvest exhausts the trees from reserve food materials leading to no flowering in the following season. If the soil provides plenty of nutrients like nitrogen, the tree develops an excess of vegetative growth that will delay the growth of fruiting buds. The role of C/N ratio for floral initiation has often been investigated by measuring levels of stored carbohydrates or imposing treatments such as fruit thinning, girdling that modify the levels of stored carbohydrates, and correlating these with flowering intensity in many woody perennials. Urea enhances initiation of citrus flowering. In Strawberry number of flowers and inflorescences increased when plants were applied with potassium nitrate. Foliar spray of potassium, ammonium, or calcium nitrate stimulates flowering of mango in the low-latitude tropics. Regular soil and plant tissue nutrient diagnostics and timely supply of nutrients in right quantity is essential to maintain health of the tree and balance between vegetative and reproductive phase. Adequate levels of minerals are important and frequent feeding through use of fertigation and/ or post-bloom nutritional sprays appear to be helpful.

f) Managing temperature

Some cultivars flower more reliably than others and will flower at higher temperatures. Floral promoter in apple is dependent upon cool temperatures of spring to enable floral induction at the appropriate time, as in the case of tropical plants. When a plant of mango, citrus or litchi is exposed to warm temperatures (30° C day/25°C night) at the time of shoot initiation, the resulting shoot growth is purely vegetative and if maintained in cool conditions (18 °C day/10 °C night), it produces generative shoots. Vegetative or generative shoots are evoked according to conditions present at the time of initiation. Vegetative (V) or generative (G) shoot types can be reversed in litchi and mango during shoot morphogenesis. Transition shoots (V>G or G>V) were evoked when containerized trees were transferred from warm-to-cool or cool to-warm temperatures, respectively, during early bud development. Some avocado cultivars and citrus may flower more than once per year under certain subtropical conditions.

g) Deficit irrigation for flower induction

The timing and level of water stress at critical phenological stages are important for success on canopy structure, vegetative and reproductive growth, protection against winter and summer injury. Mild water stress applied during the period of slow fruit growth control excessive vegetative growth while maintaining or even increasing yields. Moderate and severe deficit irrigation advances flowering date while light water-stress caused reduction in watering along the season, failed to modify flowering date. The hypothesis is that a moderate water stress switches on the flowering program without restraining panicle development. In several species of genus Citrus, a severe water stress in summer initiates a second bloom after re watering and results in a more valuable crop next summer. In mango, water stress also advance bloom date, but is not essential for induction of floral morphogenesis and cool temperatures have been identified as the main flowering stimulus. In tropics, night temperatures remain too high for induction and a dry period is proposed as the environmental cue for flower induction. This tree response to deficit irrigation has been successfully exploited to induce out of season blooming and to increase the levels of flowering in many tropical and subtropical fruit crops.

Conclusion

Floral initiation and flowering in perennial fruit trees is most important event for the reproductive success and crop productivity. For successful fruit production key strategy is ability to control the timing of flowering. Fruit crops can be classified into deciduous and evergreen fruit trees. The salient features and differences between the phenologies of these two groups are necessary to understand the problems of flowering and for suggesting possible horticultural interventions.

References

Chalmers, D. J., Mitchell, P. D. & Jerie, P. H, 1984. "The physiology of growth control of peach and pear trees using reduced irrigation", *Acta Horticulturae,* 146,pp 143-149.

Cuevas, J., Cañete, M.L., Pinillos, V., Zapata, A.J., Fernández, M.D., González, M., Hueso, J.J., 2008. "Optimal dates for regulated deficit irrigation in „Algerie loquat (*Eriobotrya japonica* Lindl.) cultivated in southeast of Spain", *Agric. Water Manage,* 89, pp 131–136.

Davenport, T. L., 2006. "Pruning strategies to maximize tropical mango production from the time of planting to restoration of old orchards" *HortScience,* 41,pp 544–548.

Davenport, T.L., 2010. "Mango: reproductive physiology", in: DaMatta, F., (ed.), Ecophysiology of Tropical Tree Crops, Nova Science Publishers, Inc. New York. pp. 217-234. Erika Varkonyi-Gasic, R. Wu, S. Moss and R.P. Hellens, "Genetic Regulation of Flowering in Kiwifruit" *in Proc. VIIth IS on Kiwifruit. Acta Hort.* 913, 2011, 221-227.

Khayat M, Rajaee S, Shayesteh M, Sajadinia A, and Moradinezhad F, 2010. "Effect of potassium nitrate on breaking and dormancy in strawberry plants", *J Plant Nutr,* 33(11): 1605-1611.

Lionakis S. M and Schwabe W. W, 1984. "Bud dormancy in the kiwi fruit, Actinidia chinensis", *Annals of Botany,* 54, pp 467–484.

Nunez-Elisea, R and Davenport, T. L, "Effect of leaf age, duration of cool temperature treatment, and photoperiod on bud dormancy release and floral initiation in mango", *Scientia Hort.* 1995, 62, pp 63–73.

Ramírez, F., Davenport, T. L., Fischer, G., 2010. "The number of leaves required for floral induction and translocation if the florogenic promoter in mango (*Mangifera indica* L.) in a tropical climate". *Sci. Hortic.* 123,pp 443–453.

8

Factors Affecting Fruit Set, Development and Quality

Nirmal Sharma and Amit Kumar*

Division of Fruit Science, SKUAST-Jammu, Jammu and Kashmir, India
Division of Fruit Science, SKUAST-Kashmir, Jammu and Kashmir, India

Fruit is a structure which arises from an ovary or fusion of several ovaries after fertilization and may or may not carry associated floral parts. Fruit formation is an evolutionary mechanism designed to support and dispersal of seed by various agencies. With few exceptions e.g. banana, it is necessary for the flowers to be fertilized to set fruit. Fruit set succeeds the process which involves fusion of male and female gametes inside the ovary and is called fertilization. Fruit setting has long been remained a major problem as is evident from the old Greek literature and wall paintings where male panicles of date palm were being rubbed against the female flowers to get good fruit set. Lack of pollination with compatible source leads to the flowers drop except the flowers which set parthenocarpically. Further if fertilization is affected even then the flower drop and this decision depends upon the internal hormonal levels of the plant and prevailing climatic conditions. On the basis of pollination requirement fruit plants can be autogamous, allogamous or geitonogamous. In case of fruit plants like grape, peach where pollen produced by the same flower affect the fertilization are called autogamous and fruits like apple, kiwi, pistachio necessarily require pollen produced on some other plant or cultivar which are compatible can affect fertilization are called allogamous. Another condition exists where pollen grains from the same plant affect fertilization in flower at different part of the plant e.g. Walnut, pecan nut and is called geitonogamy. Similarly on the basis of compatibility between pollen and stigma, plants can be self unfruitfull, partially self fruitfull and self fruitfull. Self fruitfulness indicates that fruit species or cultivars produce abundant viable pollen grains whose shedding coincides with the receptivity of stigma, pollination is effective and there are no pre and post fertilization barriers. However, partial self fruitfulness and self un-fruitfulness exhibits number of causes. In case of fully self fruitfull types, a solid block of a single cultivar can be raised without making any provision for cross pollination, however it has been reported that provision of cross pollination in autoganmous crops increases the fruit-set. In case of self unfruitful and partially self unfruitful types, provision of cross pollination is necessary to get the fruits. Proper selection of the cultivars is therefore important from the point of fruit yield for which besides overlapping of blossoming season over long periods, the time of maturity, productiveness, market values etc. are of importance. Pollenizer must be a heavy pollen producer to ensure better chances of cross fertilization. Although pollenizers are desirable, their number must be restricted so that these do not become uneconomical because the pollenizer trees occupy space, consume nutrition and take away attention of the fruit grower without any return in yield, which would have otherwise been diverted to the trees capable of bearing fruits. Numerous environmental, edaphic, plant characters affect the fruit set and are discussed as under:

1. **Pollination:** Pollination refers to the transference of viable pollen grains to the receptive stigma and is foremost pre-requisite for fruit set. In case of self pollinated fruit crops, pollination is not a problem but in case of cross pollinated crops, pollination becomes a serious concern. Usually most of the fruit plants are cross pollinated and even if they areself pollinated, provision of cross pollination improves the fruit set. Growth of the ovary by cell division and cell expansion continues to sexual maturity of the flower, but then stops at the time of anthesis or shortly before pollination. The decision to resume growth is taken only if pollination occurs. Ovaries remain receptive to pollen for some time after anthesis which varies with species, but if they are still unpollinated to the end of the period they undergo senescence and abscission. Retention of the ovary under the stimulus of pollination is known as setting of fruit or fruit set. Bees and other pollinators play an important role in orchard in providing this stimulus for fruit set. For fruit set to occur, pollination is necessary, but fertilization is not. For instance, fruit set can be accomplished by the placement of foreign, but compatible pollen on the stigma or by application of an aqueous pollen extract to the stigma or ovary. In both cases fertilization does not occur.Pollen extracts are believed to be rich in auxins, especially in cultivars that are amenable to parthenocarpy. It appears that pollination either provides the stimulus for the young ovary to synthesize its own endogenous auxins, which in turn promotes its growth and also inhibits its abscission.Pollen grains of many species are also rich in brassinosteroids, but the effects of brassinosteroids on fruit set has not been investigated.

2. **Pollinator:** Pollinator refers to the pollinating agency which carries pollen from anther to the stigma. Insects and wind are two most important

pollinators in fruit crops. Wind pollinated types are called anemophilous and insect pollinated are called entomophilous. Among insects, honey bees who visit the flower for nectar and pollen grains affect pollination in many fruit crops, though house house flies does affect pollination in mango. In case of cross pollinated species, pollinizer and pollinator are of utmost importance, otherwise nothing is produced. If adequate provision of pollinizer and pollinator is made, fruits rae provided other factors remain favorable. Therefore it is always advised to keep honey bee boxes in the cross pollinated crops. The number of bee boxes depend upon orchard area, age and size of the tree, climatic condition etc. Though provision of bee colonies is necessary for fruit set in cross pollinated plants, but literature shows that fruit set improves in self pollinated crops after provision of bee colonies is made in the orchard.

3. **Pollinizers:** Sometime certain cultivars produce non-viable pollen grains (sterility) or pollen is not able to affect fertilization (incompatibility). Dushehri, Langra and Chausa mango are self incompatible. Most apple cultivars are self incompatible. JH Hale peach and Bartlett and Kieffer pear are self sterile. Under such situations another pollen source/ cultivarinterplanted in the orchard which produces viable pollen and is compatible with the main cultivar can cuase fruit set. Thus adequate provision of pollinizer is necessary for making the main cultivar fruitful. Provision of pollinizer can be done by different methods i.e. interplanting, bouquet method, grafting method etc.

4. **Sex distribution:** On the basis of presence and absence of sex, plants can be monoecious or dioecious. Only one sex is present on a plant in case dioeciousness (date palm, papaya, kiwi, pistachio and muscadine grapes) and both in case of monoecious, both the sexes are present on a plant. In case of dioecious plants suitable proportion of male and female plants has to be made to get good fruit set. The solid block of a single type will be barren.

5. **Dichogamy:** Dichogamy refers to the maturation of male and female organs at different time. Protandrous refers to the condition when anther shed pollen grains and stigma has not yet become receptive, whereas protogynous refers to when stigma has become receptive but anthers has not released pollen grains. Non-synchronization of the maturity timing of plant sex organs leads to unfruitfulness provided nay other compatible pollen source is not arranged at the time of stigma receptivity.

6. **Sex ratio:** In case plants which produce hermaphrodites flowers, male and hermaphrodite flower (andromonoecious), or male and female flowers,

the fruit set is determined by the percentage of hermaphrodite/female flowers on the plant e.g. in case of mango and aonla which produces male and hermaphrodite flower and walnut and pecan nut which produces male and female flowers.

7. **Defective sex organs:** Sometimes in healthy looking flowers the sex organs i.e. anthers or pistils are defective thus leads to unfruitfulness.

8. **Duration of stigma receptivity:** Usually in most fruit crops the stigma starts becoming receptive one day prior to anthesis and continues two or more days after anthesis with peak receptivity on the day of anthesis. Broader the stage of stigma receptivity better will be the fruit set.

9. **Environmental factors:** Some of environmental factors do also affect fruit set.

   Moisture stress: Deficit and excess of moisture at the time of fruit set are harmful. Moisture deficit is more harmful as it leads to flower shedding post fertilization. The excessive atmospheric moisture or rainfall limits the pollination and pollinator activity thus pollens may not be available for pollination at the proper stage.

   Temperature: Low temperature stops the activity of pollinators and very low temperature kills the developing flower buds. High temperature also limits the pollinator activity and may lead to desiccation and flower drop. High temperature may also cause non viability of pollen grains and stigma surface dries at quiet faster rate which lowers down the effective pollination and fertilization period.

10. **Hormonal factors:** Post fertilization period triggers number of biochemical processes inside the ovary which produces number of hormones. Lack or excess or these hormones could lead to fruit drop.Low auxinconcentration has been found to be associated with excessive fruit drop.The postulate that pollination and fertilization induce fruit growth partially independently of auxin is also supported by the differences in fruit growth between pollinated and auxin-induced fruit. So far, it is unclear where the first auxin is produced after pollination, or whether it is transported to other tissues of the ovary.

11. **Leaf number:** Leaves capture light energy and use it to synthesize reduced carbon compounds from carbondioxide and water in a process called photosynthesis. Photosynthesis produces carbohydrates and plant growth and cropping depend on a ready supply of carbohydrates and nutrients (Oliveira and Priestley, 1988). Leaf drop due tostress greatly influence leaf-fruit sink-source relationships and can be caused by

environmental air or root conditions e.g. temperature, drought, salinity, oxygen deficiency (Fischer, 2011). In flooded trees, leaves dropless when they are in full fruit development, as compared to those in other physiological stages (Lenz, 2009).Carbohydrates are removed with fruits during harvest and the leaves are the organs of high carbon uptake by the plant. After harvest, all practices that favour carbon uptake such aslight and health should be optimized (Lenz,2009). Fruit removal in apple trees favours more leaf area development compared to those with intact fruits (Lenz, 2009) and subsequent fruiting in young trees reduces leaf area. Defoliating trees partially increases the rate of photosynthesis in the remaining leaves because they provide a relatively larger sink (Kozlowski and Pallardy, 1997) and this depends on the defoliation degree. Leaf removal on citrus at the beginning of fruit cell division causes fruit abscission which increases with increasing defoliation (Agustí, 2004). Minimum quantities of leaf area and shoot structure are required for setting large fruit crops (Lakso and Flore, 2003). Theleaf area index in conjunction with sunlight interception is useful as a basis for analyzing canopy productivity (Fischer, 2011). Leaf area index in the apple lies between 1.5 and 5 depending on the variety, rootstock, pruning, trellising, fertilization and other cultural practices Jackson (1980). The index in the peach is generally higher, between 7 and 10 (Faust, 1989). Moreover, height and type of training define light penetration to the foliage (Faust, 1989). The leaf area index in orange canbe as high as 9 or 11 (Dussi, 2007).

Fruit Development

Botanically fruit is ripened ovary; however horticultural fruit is ripened ovary with or without the accessory parts which supports the seeds. The initiation of fruit development starts with the trigger of fertilization or the stimulus of fertilization. Fruit, from fruit set to maturity takes about 5-6 months. This fruit development period involves lot of physical and physiological changes. Usually the fruit development exhibits sigmoid growth curve which can be divided into following four phases: Post fruit-set period involves fruit development period and is divided into different stages. In most species the fruit growth can be represented by single sigmoid growth curve or double sigmoid growth curves (peach, plum, apricot, current and seeded grape), however triple sigmoid growth curves(kiwi) are also observed. In double sigmoid curves there is second burst of growth during the ripening period. Physiologically and biochemically, fruit development can be divided into 4 phases, which although continues, are separated on the basis of the major activities.

1. It is the initial phase of the fruit development which includes ovary development, anthesis and dehiscence of anthers. It involves slow division of cells.

2. This is the post fertilization period when fruit starts growing rapidly due to rapid cell division.

3. In this phase cell division almost ceases and the fruit growth is attributed to cell enlargement. Food reserves are accumulated and fruit attain their final shape and size.

4. This is the final phase of fruit development and involves ripening of the fruit which is followed by shedding.

In case of stone fruits the first slow growth period involves lignifications of the endocarp which leads to pit hardening. Cell division usually ceases in phase-3, however in fruits like avocado and strawberry, cell division continues in phase three also. The development of the fruit size depends on anumber of factors such as the leaf-fruit ratio, geneticand climatic factors, position in the plantand the branch, tree age, number of seeds, water and nutrient supply (Dennis, 1996). Fora full crop, most fruit species will set more fruitthan needed if growing conditions are optimal (Westwood, 1993) which are later shed during different fruit drop waves if required. During their development, fruits accumulatecarbohydrates, generally as starch, sucrose, orhexose sugars (Kozlowski and Pallardy, 1997) which are highly dependent on the fruit maturity stage and varies according to cultivar, leaf : fruitratio and growing conditions (Friedrich and Fischer, 2000). The breakdown of starch inmango fruit mainly leads to an increase in sucrosecontent rather than in glucose (Léchaudel and Joas, 2007).The fruit attracts photosynthates and thus increasesthe photosynthetic production of leaves(Kozlowski and Pallardy, 1997), while few fruitsin the canopy cause accumulation of photosynthatesin leaves (less photosynthesis activity) (Hansen, 1982). A high fruit load can induce vegetative growth stagnation (Kozlowski and Pallardy, 1997). During full fructification, over 80% of the photosynthates can be used for fruitfilling (Schumacher, 1989). While a high fruit load decreases the distribution of assimilates to the roots and other permanent plant organs, the lack of assimilates may also have negative effects on fruit production in the followingyears (Lenz, 2009). Poor accumulation of reserves in persimmon (*Diospyros kaki*) inhibited flower induction, causing alternate bearing (Ojima*et al.*, 1985). This same phenomenon has alsobeen reported in the literature for species suchas citrus (Goldschmidt and Golomb, 1982) and the apple (Lenz, 2009). Consequently, vegetative growth must be sufficiently vigorous to enable growth of well-illuminated leaves (Gil, 2006). Fruits show a strong attraction for photosynthetic products and if the amount of fruit rises, the photosynthate production by leaves is higher (Kozlowski and Pallardy, 1997). Expanded leaves near the fruit exhibit increased photosynthetic rates (Urban *et al.*, 2003). Hansen (1978) observed that the distant leaves can serve as an assimilate source and thus, the importance of having more fruits in the thinned branches. Green and immature fruits exhibit substantial surface-to-volume ratios and refix a lot of internally respired carbon but only modest amounts of atmospheric CO_2, mainly during early development (Blanke and Lenz, 1989).

Hormonal Changes During Fruit Development

Three hormones i.e. auxins, gibbrellins and cytokinins are involved in the early stages of fruit growth especially phase I and II. In phase I the source of these hormones is maternal tissue, but in phase II and III which coincides with embryo growth and maturation respectively, the source of the hormones is debatable. Fruit is a maternal tissue and these hormones could be synthesized *in situ* or translocated from other parts of the plant via the phloem stream. There are also instance where fruit growth is affected by auxins and or cytokinins produced by the embryo and endosperm. During fruit development, two peaks of auxin are observed. The first peak reaches its maximum 8 days after pollination at the end of active cell division, and the second peak reaches its maximum at 30 days after pollination. The latter was not found in parthenocarpic fruit (Mapelli *et al.*, 1978), suggesting that, at least during the later stages of fruit development, the embryo supplies the auxin necessary for continued fruit growth. The observations that parthenocarpic fruit are generally smaller than wild-type fruit (Sjut and Bangerth, 1983) and that there is a positive correlation between final fruit size and number of seed in the fruit (Varga and Bruinsma, 1976) support this hypothesis. Lemaire-Chamley *et al.* (2005) showed that candidate key genes for auxin biosynthesis, transport, signaling, and responses were already expressed in the locular tissue during the early stages of fruit growth. More detailed analysis of genes differentially expressed between the locular tissue and the outer pericarp, revealed that the expressions of these genes follow a gradient from the central part of the fruit (placenta and locular tissue) to the outer part of the fruit (Lemaire-Chamley *et al.*, 2005). It is possible that, in response to pollination and fertilization, the auxin is newly synthesized or hydrolyzed from its conjugates in the central parts of the fruit, the developing ovule and/or its surrounding tissues, respectively, and subsequently transported to the outer layers. This transport leads to the formation of the auxin gradient in the fruit tissues, which is translated into auxin responses, such as cell division, cell expansion, and into cross-talk with other hormones, such as newly synthesized gibberellins (Lemaire-Chamley *et al.*, 2005). Phase II and III are marked by rapid cell division and rapid cell expansion respectively. These periods coincide with the growth of the embryo and endosperm, and it has been suggested that auxins and cytokinins produced by the embryo and endosperm have a role in fruit growth.

Factors Affecting Fruit Development and Quality

1. *Pollination:* Although successful pollination causes fruit set and some growth of young fruits, a second stimulus in the form of fertilization provides an impetus for continued growth under phase II. Fertilization is not the only signal that affects fruit growth or retention. Competition among different metabolic sinks for photo-assimilates and resource allocations at the whole plant level are involved. Fruit drop occurs despite successful fertilization and embryo formation and determines what proportion of the fruit crop is to be retained up to harvest. A direct influence of pollen on fruit size and quality is called metaxenia. One of the few examples of this is the date where pollen source affects the sugar content and size of fruit.

2. *Number of seeds:* A way to decide, whether to retain or drop fruit is through number of seeds. Effect of number of seeds on fruit growth has been practically demonstrated in number of fruits like strawberry, apple and grapes etc.Exogenous application of NAA, a synthetic auxin to the deachened receptacle in strawberry could replace the stimulus of fertilization for fruit growth which suggests that young seeds with developing embryos and endosperms could be a rich source of auxins and this auxinis the trigger for continued growth of the fruit. Many plants produce fruits that either lack seeds or have no viable seeds. The production of such seedless fruits is known as parthenocarpy e.g. banana, pineapple, cultivars of grape, fig, oranges grapes and kiwi etc. In natural populations parthenocarpic fruits result from one of three causes: i) lack of pollination, ii) pollination occurs but not fertilization and iii) Embryo abortion preceeding fertilization. Navel oranges and Thompson seedless show seed abortion. Hormones, auxins, gibberalins and cytokinins especially the first two are well known to induce parthenocarpy. Thus auxin treatment of young unpollinated ovaries in certain species is known to cause production of parthenocarpic fruits. Different fruit crops vary in seed content, peach having one seed, and tomato have numerous seeds. Most commercially seedless varieties of orange ('Hamlin', 'Valencia') and grapefruit (Marsh and Redblush) have 0-6 seeds.Sindhu mango does have stone but it is too thin. The lack of seeds in Tahiti lime is due to sterile (nonfunctional) pollen and ovules. Commercial seedlessness in Hamlin and Valencia oranges and grapefruit is due to very low level of

fertile ovules. Marsh and Redblush have low levels of pollen and ovule fertility. Since high levels of ovule (female) sterility is common in all commercially seedless oranges and grapefruit, there is no way to increase seediness appreciably. Even though placing extremely large quantities of viable pollen by hand on some commercially seedless varieties, such as 'Valencia' oranges, will slightly increase seed content It is important because fruit quality of commercially seedless fruit is not reduced by a higher seed content when grown adjacent to varieties with large amounts of viable pollen. Several mandarin hybrids like Orlando, Minneola, Robinson, Nova and Sunburst are sexually self-incompatible. Such varieties have viable pollen and fertile ovules. There is a chemical inhibitor in the style, however, that reduces pollen tube growth so much the style abscises falls off before the pollen tube can enter the ovary. Therefore, no seeds are produced. Since sexual fertilization is normally needed for fruit set, such varieties should theoretically not set fruit. Some varieties, however, have the capacity for setting fruit without sexual fertilization termed parthenocarpy. Weak parthenocarpic fruits fall under conditions of physiological stress, primarily drought that would not cause seedy fruit to drop. A mandarin hybrid, Page, is strongly parthenocarpic and tolerates more stress without shedding fruit. By providing pollinizer varieties with sexually compatible, pollen good yields are obtained, however, the resulting fruit is seedy and fresh fruit quality is thereby reduced.

3. *Number and divisions of cells:* Number of cells in the mature unfertilized ovary and subsequently in early fruit growth is a critical parameter for the final fruit size. Cytokinins play an important role in regulating cell division. The cultivar Hass of avocado produces two populations of fruit, one significantly smaller than the normal one. The cell size in the two phenotypes is the same but, the smaller phenotype results from a reduced number of cell divisions which in turn is related to reducecytokinin/ABA ratio in the smaller compared to the normal phenotypes. The deficiency is corrected and the fruit size is restored to normal size by an exogenous application of cytokinins, Isopentyladenine, in phase I. Significantly the deficiency is not corrected by the application of gibberelin. Studies correlating gibberelin content with fruit growth are relatively few but in number of fruit plants the fruit and seed growth have been correlated with endogenous gibberellin content.ABA negatively regulates fruit growth, probably by inhibiting both cell division and cell growth.Gibberelic acid has been used to make grape berries longer and bigger and pedicel also gets elongated making berry separation easy and adds to the table appeal but high crop load is probably the main cause of alternate bearing (Iglesias *et al.*, 2007). Photosynthate production is often unable to satisfy

the demandsduring fruit set and fruit growth following heavy and prolonged flowering (Chackoet al., 1982).

4. **Phytohormones:** The influence of seed on fruit growth was shown by Nitsch (1950) in strawberry. Nitsch removed varying number of developing achene from young fruit of strwawberry and showed a strong correlation between the number of achene left and the fruit growth around them. Similar correlation between number seeds and fruit growth have been shown in other fruits like grapes, blackberry and apple etc. He also concluded that young seeds with developing embryos and endosperm are rich source of auxins and that auxins are trigger for the growth of the fruit.Auxins regulate almost all developmental processes in plants, including fruit development. Natural and artificial auxins supplied exogenously to unpollinated flowers induce fruit growth in horticultural plants, suggesting that these hormones can replace the signals provided by pollination and fertilization (Nitsch, 1952 and Schwabe and Mills, 1981). This hypothesis is in accordance with the finding that increased auxin levels are detected in flower organs after fertilization of the ovules (Gillespyet. al., 1993). Recently, molecular analyses have confirmed the prominent role played by auxinsignaling in triggering and coordinating the transition from flower to fruit (Vriezenet. al., 2008, Wang et. al. 2009 andDorceyet. al. 2009). Auxin is synthesised in meristem and young leaves and transported in a polar fashion to the other parts of the plant (Leiser, 2006).In addition to auxins, cytokinins and gibbrellins also play a role in the growth of the young fruits. Cytokinin level especially zeatin and zeatinriboside are high at earliest stage of fruit growth and has been correlated with the phase of maximum cell division. In kiwi fruit, the concentration of total cytokinins was high at the earliest time after anthesis, dropped to low levels later.In a non-climacteric fruit kiwi, the concentration of total cytokines is high at the earliest time after anthesis, drops to low levels and rise again before harvest 175 days after anthesis. Date for zeatin and isopentyladenineare generally similar. Gibberellins are another group of phytohormones that plays a prominent role in coordinating fruit growth and seed development. Gibberellins are able to induce fruit set in several horticultural species. However, gibberellins induced fruits are smaller than seeded fruits suggesting that other signals are required for fruit growth and development (DeJong et. al. 2009). Increased levels of gibberellin have been detected in pollinated ovary together with an increased expression of gibbrellin biosynthetic genes (Serrani et. al. 2007). On the other hand, the application of inhibitors of gibbrellin biosynthesis limits fruit growth (Serraniet. al. 2008). A synergistic effect of auxin and gibberellin on fruit growth has been observed, suggesting that the two

phytohormones interact in regulating fruit development (Serrani*et. al.* 2007 and Serrani*et. al.* 2008). Therefore it is established that auxins, GAs and cytokinins are involved in the early stages of fruit growth especially in phase I and II. Endogenous level of IAA shows highest level at the earliest time of fruit growth. This level then drops steadily and again rises at the time of ripening.

5. **Photo-assimilates:** During their rapid growth in phase II and III, fruit act as strong sinks and import massive amounts of photoassimilates from leaves. Such translocated material is mostly sucrose, although in some species, oligosaccharides (raffinose) or hexitols (mannitol, sorbitol) may be the predominant sugar/sugar alcohol. On arrival, the sugar (sugar alcohol) may be converted to starch (mango, banana and kiwifruit) or stored as reducing sugars (tomato, strawberry) or stored as sugars (muskmelon, watermelon and grape). In some fruits it may be converted to lipids e.g. olives. Many of this stored food undergo further modifications during ripening. Various enzymes involved in sugar-starch metabolism e.g. acid invertases (for hydrolysis of sucrose to hexoses, fructose and glucose), starch synthase and starch branching enzymes are active in growing fruits. By removing sucrose from the site of unloading, these enzymes play an important role in maintenance of the sucrose concentration gradient in the phloem stream and in the sink strength of the fruit.During the peiord of rapid growth in phase 2 and phase 3, fruits act as a strong sink and import massive amount of photo assimilates from photosynthesizing organs. Such translocations occurs in the phloem tissue and the translocated materials are mostly sucrose, although in some species oligosaccharides (rafinose) or hexitols (maintol and sorbitol) may be the predominant sugars/sugar alcohol. Hormones have been implicated in regulating the partitioning of photoassimilates between competing sinks. In many studies ABA has been suggested as the hormone that facilitates unloading of the photoassimilates at the sink site, but evidence are equivocal.

6. *Leaf number:* Plant growth and cropping depend on a ready supply of carbohydrates and nutrients (Oliveira and Priestley, 1988) and photosynthesis produces carbohydrates (Lakso and Flore, 2003). Leaves accumulate carbohydrate at a high rate, *e.g.*, 8 mg/g/day of dry weight in citrus (Kozlowski and Pallardy, 1997).Young leaves depend in part on the carbohydrates imported from other areas of the plant, whereas, mature leaves produce excess photosynthates and act as the principal source of the plant of translocated carbohydrates (Turgeon, 1989) which are stored in the fruits. Because carbohydrates are removed with fruits during harvest

and the leaves are the organs of high carbon uptake by the plant, after harvest, all practices that flavour carbon uptake such as light and health should be optimized (Lenz, 2009). Fischer *et. al.* (2010) recommended maintaining peach trees with intact leaves 3-4 months after harvest, before defoliation, to improve carbohydrate accumulation for the next cycle. Defoliating trees partially increases the rate of photosynthesis in the remaining leaves because they provide a relatively larger sink (Kozlowski and Pallardy, 1997). Leaf removal on citrus at the beginning of fruit cell division cause fruit abscission which increase with increasing defoliation (Agustí, 2004). The fraction of the average light passing through the canopy is diffuse non-interceptance (DNI) ranges from 0.02 to 0.36, maximum being 1). The varieties with a low leaf area index and high diffuse non-interceptanceare better exposed to solar radiation and produce more reproductive stems and good colour fruits than varieties with denser foliage (Rajan*et al.*, 2001). Apart from cultural practices, agro-ecological conditions and age of plants can influence leaf area index development (Fischer, 2011). The optimum exposure of the maximum number of leaves to light normally results in the greatest yield of dry matter (DeJong and Ryugo, 1998). Optimal leaftofruit ratio varies according to the species and variety, and orchard geographic location (Schumacher, 1989). Moreover, the capacity of leaf photosynthesis depends on the incidence of light, whereby the shaded parts of the canopy assimilate less and need more leaves than the well illuminated part for optimal fruit development. Schumacher (1989) considered that the leaf-fruit ratio is not totally reliable. Hansen (1978) stated that decreasing the leaf/fruit ratio increases the photosynthetic efficiency of the leaves, causing a raised sink-effect. Tree fruits with a high leaftofruit ratio, as inyoung plants or those with a low fruit load, often form large fruits with a "spongy" tissue which reduces postharvest life and increases susceptibility to diseases (Fischer and Friedrich, 2000). As fruit density increases, the leaftofruit ratio decreases, resulting in a lower supply of photosynthate per fruit therefore fruit size decreases (Dennis, 1996), along with insufficient colour and flavour (Schumacher, 1989). Optimal leaf area in several fruit species is 200 cm^2 per 100 g of fresh fruit mass for favourable growth and quality (Fischer, 2011). Furthermore, grapes require twice this value. The increase in leaf-fruit ratio may facilitate the accumulation of starch reserves, flavouring vegetative growth and fruiting in the following season (Chacko *et al.*, 1982). Grapevines doubled the root starch concentration from 12 to 25% dry weight when the leaf-fruit ratio increased from 0.5 to 2.0 m^2 of lightexposed leaf area per kg fruit (Zuffereyet *al.*, 2012). The rate of fruit sucrose accumulation in Satsuma mandarin is higher at a normal load (25 leaves/fruit), as compared to trees with 50 leaves per fruit (thinning at 70 days after anthesis) (Kubo *et al.*, 2001). Thinning of 10, 25, 50, 100 and 150 leaves per fruit in the mango cultivar Lirfaresulted in the highest fresh weight of fruit at 100 leaves, while flesh dry weight increased 11%, when the number of leaves increased from 10 to 100 (Léchaudel *et al.*, 2004). The leaf-fruit ratio changes with the production area latitude in which the temperature and light have the greatest influence, with lower ratios at sites nearer to the equator (Fischer, 2011). Sauer and Baumann (2007) reported that for the production of 1 g of grape berries, 20 cm^2 of leaf area are needed (for 1kg, $2m^2$ leaf area).As fruit density increases, the leaf to fruit ratio decreases, resulting in a lower supply of photosynthate per fruit; fruit size therefore decreases (Dennis, 1996) along with insufficient colour and flavour (Schumacher, 1989). The increase in leaf-fruit ratio may facilitate the accumulation of starch reserves, flavouring vegetative growth and fruiting in the following season (Chacko *et al.*, 1982).

7. *Cultural operations:* The capacity of leaf photosynthesis depends on the incidence of light, whereby the shaded parts of the canopy assimilate less and need more leaves than the well illuminated part for optimal fruit development.Casierra-Posada *et al.* (2007) observed increased TSS content and pulp/stone ratio in the Rubidoux peach with thinning, the optimal fruit quality being at 40-50 leaves/fruit. Growers can rely on a number of methods which directly or indirectly influence photosynthesis and sink activity (fruit growth). Among these, the most important are tree height, distance, fruit thinning, pruning, fertilization, application of growth regulators, irrigation and phytosanitary control (Flore and Lakso, 1989; Fischer, 2005). Girdling involves the removal of a bark ring in the trunk or in the base of lateral growth axes, interrupts photosynthate flow to theroots and thereby increases flower induction and fruit filling, apparently through increased sugar availability in the aerial parts of the tree (Iglesias *et al.*, 2007). Ringing (5 mm wide) the base of productive main branches of Sweet Orange trees, 3 weeks after anthesis, increases fruit retention by 38%, as compared to non-ringed trees (Cabezas- Gutiérrez and Rodríguez, 2010). Girdling trees greatly increases the fruit size and is commonly practiced in grape but usually only vigorous trees, such as relatively young ones, respond well to this treatment. Proper time of girdling is from full bloom to two-thirds petal fall but some success has been obtained even at complete petal fall. Gibberellic Acid applied at the rate of 10-15 ppm in full bloom increases yield of seedless fruits of several self-incompatible varieties. Applications are made as an aqueous spray that wets the canopy well. At higher concentrations leaf drop occurs but only the older leaves

fall. Training and pruning alters the balance between vegetative growth and reproductive fruiting by the allocation of resources, such as carbohydrates, water and growth regulators (Myers, 2003). Heavy pruning diminishes leaf area, whole tree photosynthesis and translocation of photosynthates to fruits and roots, increasing the root/shoot ratio (Casierra-Posada and Fischer, 2012) and favouring vegetative growth. In guava, mid- and light pruning provide greater fruit weight ratios in contrast to heavy pruning (Serrano *et. al.*, 2007). Duringthe reproductive phase fruiting pruning is used because this pruning type improves fruit load, regulates the physiological balance, ensures a harmonious and rational distribution of high quality production, maintains a constant production over time, and contributes to fruit thinning (Arjona and Santinoni, 2007). In pruning, it is important to cut off (thinning) upright water sprouts which direct photosynthates, among other substances, to the growing shoot tip at the expense of reproductive growth (Myers, 2003).Heavy pruning diminishes leaf area, whole tree photosynthesis and translocation of photosynthates to fruits and roots, increasing the root/shoot ratio (Casierra-Posada and Fischer, 2012) and favouring vegetative growth.

8. *Changes during ripening:* Fruit ripening is accompanied with change in the fruit colour, accumulation of sugars, softening of the flesh and peel and physiologically associated with rise in ethylene production. In addition to change in sweetness, flavouring components such as organic acids (citric, mallic) and volatiles, increase during ripening and combine to produce the unique flavor and aroma of the ripe fruit. Phenolics, such as tannins, provide astringency to unripe fruit and have an important influence on the flavor and colour or mature fruit. Ripening of non-climetcteric fruits also include loss of chlorophyll accompanied by accumulation of anthocyanins, sucrose, hexoses and flavouring volatiles such as alcohols, aldehydes and their esters and softening of the pulp. It has been observed that auxins retard or inhibit ripening related changes in strawberry. Certain enzymes disrupt the hemicelluloses-cellulose network, as well as those that disrupt the pectin network. Genes for these enzymes have been cloned and described from many fruits e.g. peach, banana, passion fruit and avocado. Egases are a large group of enzymes expressed in a tissue and development specific manner. Egases expressed during fruit ripening are similar in some ways to those induced during organ abscission, both are induced by ethylene and are downregulated by auxin. Enzymes which disrupt cell wall architecture in ripening fruits are expansions, Xyloglucanendotransglycosylases.Endo-1,4-β-glucanases (EGases and cellulases) ans α and β galactosiidases. The enzymes responsible for

disruption of pectin network are polygalacturonases and pectin methylesterases. Ethylene does not act merely as a molecular switch which turns on the ripening process but rather a motor running continuously for ripening related process to continue. If ethylene production is stopped by the use of ethylene synthesis inhibitors or inhibitors of ethylene perception, the ripening process is slowed down considerably. During ripening process the chloroplast (green colour) changes to chromoplast (coloured pigments). Thylakiod membranes and chlorophyll pigments are broken down, and there is a progressive accumulation of new carotenoid pigments in the plastids. During the process of ripening, fruits show an increase in the concentration of sugars, either by hydrolysis of starch within the fruit or by continued import of sugars from other parts of the plant. The former is typical of fruits that store starch, which are often harvested before ripening (eg mango, banana and kiwi) and later fruits that ripen on vines. Moreover sucrose is hydrolyzed to hexoses, principally fructose and glucose. Activities of enzymes involved in starch/sugar metabolism, such as sucrose phosphate synthase and acid invertase have been shown to rise in several fruits and their mRNAs have been shown to be upregulated by ethylene.

9. *Fruit quality:* Fresh fruit quality involves but fruit size, rind texture and thickness, rind blemishes and color, seed content and juice quality etc. Pollination and related factors playanimportant role in fruit size, seed content and minor effects on some quality parameters. This effect is called metaxinia where the pollen influences the resulting seed and fruit and is very well understood in date and almond. The bitter almond kernel is produced if peach pollen grains fertilize the almond flower.The shape of the fruit depends upon the number of cell divisions and the directionality of growth of daughter cells in the various sectors of the fruit in phase II and III. Directionality of growth in fruits is probably regulated in the same manner as in growth of axial organs by precise orientations of microtubules and innermost cellulose fibrils, but there is little published information on this topic.To attain a particular shape, different fruits require cell to divide in particular direction. Banana requires cells to divide in longitudinal direction however pear requires altered patterns of cell division and growth in different portions of the fruit. Fruit size is highly important in fresh fruit markets. Within a variety seedless fruits are generally smaller than seedy ones. This is true even for commercially seedless varieties such as Valencia i.e. those fruits with a few seeds will have a larger size than those which are seedless. In seedy varieties, such as 'Pineapple' those fruits with the fewest seeds will on the average be smaller than those with the maximum number of seeds. The range of seeds per fruit

of 'Orlando' tangelo can be 0-45with a proper pollinizer variety. It has been established that there is a highly significant linear relationship between seeds and fruit size i.e. for every increase of a seed there is a corresponding stepwise increase in size. This has been demonstrated in several experiments. A stepwise increase of diameter occurs with unit increase of seeds. Other factors such as crop load, irrigation, tree vigor and rootstock also play an important role in determining fruit size. It is important to understand that while cross-pollination of self-incompatible varieties is important in assuring an adequate crop, it can be excessive and results in setting seedy fruits that are smaller than seedless fruits produced on a tree with a lighter crop load. Excessive cross-pollination results from use of too many highly effective pollinizer trees. Large numbers of 'Orlando' trees used as a pollinizer for 'Robinson' has at times resulted in excessive crops, smaller fruit and devastating limb breakage. It is apparent that seed development, rather than pollination or sexual fertilization itself, is responsible for the later stages of fruit enlargement. The embryos or immature seeds of peaches are killed by low temperature; the fruit remains extremely small even though it persists until maturity. Fruit shape is affected by seeds in some fruit species. For example, long tapering pear fruit called rat-tailed fruit. are formed when seedless. Rat-tailed watermelons have been shown to develop when an insufficient number of bee visits and thereby inadequate pollen results in few or no seeds in the stem end of the fruit. Apple fruits normally have 10 seeds spaced uniformly around the fruit. If seeds fail to form on one side of the fruit that side is flat due to lack of the hormonal stimulation of developing seeds. Seeds increase fruit size which in turn is related to fruit quality. Large fruit have lower total soluble solids and acid, higher ratio and more juice per fruit, but lower percent juice on a weight basis. Seeded fruits have higher total soluble solids and acid.

References

Arjona, C. and L. A., Santinoni. 2007. Poda de árbolesfrutales. pp. 243-282. In: Sozzi, G.O. (ed.). Árbolesfrutales, ecofisiología, cultivo y aprovechamiento.Universidad de Buenos Aires, Buenos Aires.

Cabezas-Gutiérrez, M. and C.A., Rodríguez.2010.Técnicashortícolasparaoptimizar el tamaño y la calidaddelnaranjo (Citrus sinensisL.). Agron.Colomb.28:55-62.

Casierra-Posada, F. andFischer, G. 2012.Poda de árbolesfrutales. In: Fischer G.(ed.) Mannualpara el cultivo de frutales en el trópico.Produmedios,Bogotá.pp 169-185.

Casierra-Posada, F., Rodríguez, JI.and Cárdenas-Hernández, J. 2007. La relaciónhoja:frutoafectala producción y la calidad del fruto en duraznero (PrunuspersicaL. Batsch, cv. Rubidoux).Rev. Fac. Nal. Agr. Medellin 60(1):3657-3669.

Chacko, E.K., Reddy, Y.T.N.andT.V., Ananthanarayanan. 1982. Studies on the relationship between leaf number and area and development in mango (MangiferaindicaL.).J. Hort. Sci. 57(4): 483-492.

De Jong M, Mariani C, and W.H., Vriezen . 2009. The role of auxin and gibberellin in tomato fruit set. *J. Exp. Bot.* 60:1523-1532.

Dennis FG. 1996. Fruit development. In:Maib KM, Andrews PL, Lang GA and MullinixK (eds.). Tree fruit physiology: growth and development.*Good Fruit Grower*.pp107-116.

Dorcey, E., Urbez, C., Blazquez, M.A., Carbonell, J., and A., Perez-Amador. 2009. Fertilization-dependent auxin response in ovules triggers fruit development through modulation of gibberellin metabolism in Arabidopsis. *Plant J.* 58:318-332.

Dussi, M.C. 2007. Intercepción y distribuciónlumínicaen agro-sistemasfrutícolas. In: Sozzi,G. (ed.). Árbolesfrutales: ecofisiología, cultivo yaprovechamiento. Editorial Facultad de Agronomía,Universidad de Buenos Aires, Buenos Aires.pp. 200-241.

Faust, M. 1989. Physiology of temperate zones fruittrees.John Wiley and Sons, New York, NY.

Fischer, G. 1995. Effect of root zone temperature and tropicalaltitude on the growth, development and fruitquality of cape gooseberry (*Physalisperuviana*L.).Ph.D. thesis.Humboldt-Universitätzu Berlin, Berlin.

Fischer G. 2005. Aspectos de la fisiologíaaplicada de losfrutalespromisorios en cultivo y poscosecha.*Rev.Comalfi*. 32(1):22-34.

Fischer G. 2011. La relaciónhoja/fruto en especiesfrutícolas.Proc. 4th Colombian Congress of Horticulture, Palmira, Colombia.pp.40-53.

Fischer G,BeranF and Ch. Ulrichs. 2008. Partitioningof non-structural carbohydrates in the fruitingcape gooseberry (*Physalisperuviana* L.) plant. In: Book of abstracts, Tropentag 2008. Stuttgart-Hohenheim, Germany.pp145.

Fischer G, Casierra-Posada F and VillamizarC. 2010.Producciónforzada de duraznero (*Prunuspersica* (L.) Batsch) en el altiplano tropical de Boyacá (Colombia). *Rev. Colomb. Cienc.Hortíc.* 4(1):19-32.

Fischer, G. and LüddersP. 1997. Developmental changesof carbohydrates in cape gooseberry (*Physalisperuviana*L.) fruits in relation to the calyx and theleaves.*Agron.Colomb.* 14(2): 95-107.

FloreJA and LaksoAN. 1989. Environmental andphysiological regulation of photosynthesis in fruitcrops.*Hort. Rev.* 36:111-157.

Friedrich G and Fischer M. 2000. PhysiologischeGrundlagendes Obstbaues. Ulmer Verlag, Stuttgart, Alemania.

George AP, NissenRJ, Collins RJ and Rasmussen TS.1995. Effects of fruit thinning, pollination andpaclobutrazol on fruit set and size of persimmon (*Diospyros kaki* L.) in subtropical Australia. *J. Hort. Sci.* 70:477–484.

9

Fruit Drop: Causes and Control

K. Usha and Nayan Deepak G.

Division of Fruits and Horticultural Technology, ICAR-IARI, New Delhi

Fruit drop is a premature shedding of fruitsbefore harvesting for commercial purpose. There are so many reasons for fruit drop like internal (Hormonal balance, morphological and genetical) and external (biotic and abiotic) factors. Fruit drop is very much serious in some fruits like apple, peach, currant, mango, citrus etc. Fruit drop may occur at various stages of fruit growth,starting right from fruit setting till its harvesting. It may be natural, environmental or pest related.Losses due to fruit drop at various stages have long been a serious threat to the fruit growers. After determining the actual cause of fruit drop, adoption of a suitable control measure can bring relief to the growers. Among different drops, pre-harvest drop is of great economic importance which can cause serious crop loss to farmer.

Kinds of Drop

There are mainly three categories of fruit drop

1. **Post blossom fruit drop:** This is the first fruit drop begins right after the flower petals fall off and primarily consists of tiny fruits that may last two to three weeks. The dropping fruits are the ones that didn't get pollinated or the sperm cells from the pollen didn't make it to the ovary. Most fruit varieties need to be cross-pollinated by bees. Lack of pollination may be due to cold or wet weather or honeybee decline, ultimately causing fruit drop.

2. **June drop:** A second drop commonly occurs in apple, pear and less frequently, cherry in late May or June when the fruits are about marble size. This is the result of the plant's inability to support the vast number of fruits that it has produced due to profuse flowering and extensive pollination.In an effort to conserve energy, the plant drops the fruits. Essentially it is a natural thinning that result from the competition between fruits for recourses. Plant drops fruits to have large, high quality and prevent limb breakage.Herefruits which contain weak or few seeds resulting from poor pollination are the first to drop.

3. **Pre harvest drop:** It is shedding of fully developed fruits just prior to harvest, more frequent in deciduous fruit crops. In apples as ripening starts, large amounts of the ripening hormone, ethylene is produced. It leads to fruit softening and the formation of an abscission zone in the stem. Ethylene stimulates the production of enzymes (cellulase and polyglacturonase) that break down the cell walls and the glue that holds cell walls together in the abscission zone of the stem, leaving the fruit connected to the tree by only the vascular strands, which are easily broken. McIntosh, an apple variety is particularly prone to preharvest fruit drop. Another major cause for pre harvest fruit drop may be attributed to several pest and disease attack more likely to occur in fruits nearing maturity. Wormy pests may cause premature ripening and fruit fall.

Causes of Fruit Drop

The role of inner agents in the drop of flowers and fruits:

1. *Non pollination*: Self pollination may fail due to various mechanisms, while cross pollination may be prevented in absence of pollinators, suitablepollenizers or due to adverse climatic conditions. Improperly pollinated fruits are more liable to fruit drop.

2. *Over pollination*: The disadvantageous effect of supernumerary pollen grains on the stigmata, when more than 10–18 pollen gains are caught by the two lobes of the stigma, the latter faded severely just the next day and about 93% of stigmata died and dried out at the third day causing the drop of the respective female flowers. The rate of fading is closely related to the amount of pollen involved. Unviable pollen causes the same as the viable one.

3. *Non fertilization:* It may be due to following reasons:

   (a) **Gametic sterility-**This phenomenon is common in polyploids containing an uneven multiple of the basic chromosome no, especially in triploids. Such pollen grains either fail to germinate or if pollen tubes are formed these usually burst easily or after reaching the ovary, they give rise to unbalanced embryo.

   (b) **Incompatibility-** Incompatibility is due to physiological reactions occurring in between the pollen tube and the style and ovarian tissues. In black currants, the post-bloom fruit drop is mainly due to the autoincompatibility of the varieties.

   (c) **Failure of double fertilization-**It prevents the formation of endosperm and thus that of the embryo. In Napoleon variety of sweet cherry preharvest fruit drop is observed caused by the degeneration of the endosperm.

(d) **Abortion of embryo-**It may arise from genetical and unfavourable nutritional conditions and usually results in shedding of fruits.Sweet cherries of late maturity shows higher rates of fruit drop, which is attributed to the abortion of the embryo. In the super early cherries, embryo abortion occurs often during the second phase of pericarp growth, which is due to the competition for resources between the seed and the pericarp.

Seed Content of Fruits

Seeds, especially their endosperm are the sites of synthesis of growth substances like auxin which inhibits abscission of fruits. Auxin absorbs not only organic substances but also influences the distribution of cytokinins to the fruit, which is an active sink of metabolites.

The most important precondition for the fruit to be maintained on the tree is its seed content. Fruit species producing more than one seed (apple, pear, quince or currants) drop preferably those fruits. Therefore the varieties, which develop less seeds, are more susceptible to environmental adversities, *i.e.* water stress, poor nutrition, etc. and are prone to drop fruits. In apple, fruits containing less than 3 seeds are shed first when fruit set was abundant.

In one-seeded fruits – e.g. stone fruits – the seed content does not allow alternatives except yes and no in fertilisation. Fruit drop is timed by the development of the embryo(s) of the single stone. In normal fruits, the embryo grows continuously, whereas in dropped fruits, the size of the embryo stopped growing around one third of the normal size.

Competition Between the Organs of Plants

(a) Competition between the vegetative and generative organs

The relation between growth of shoots and fruit set is continuously changing during the growing season. If the source and consequently the transport *i.e.* photosynthates are restricted the fruits will drop. Balanced ratio of leaves and fruits leads for retention of fruits on the tree. At the time of the shed of petals, in apple 1–4 leaves are needed by one fruit set, around the June drop 10–15 leaves and at the end of fruit development 40 leaves provide the fruits.

Higher leaf area is necessary for fruit set more than one per inflorescence and fruit drop did not occur.A low leaf/flower ratio reduces the chances of fruit set and induces fruit drop.

Trees of strong shoot growth used to drop more fruit than the weak growing trees,which may keep often supernumerary fruit primordia on the fruiting structures.

An interesting observation that leaves may stimulate fruit abscission by influencing the translocation of ABA from the leaves to the fruits. At the same time, young leaves may delay the abscission of ripe fruits, whereas mature leaves promote the abscission of fruits by stimulating the transport of ABA.

(b) Competition between the generative organs

In a tree having large mass of flowers or fruit primordia, the distribution of food is not optimum causing a vigorous drop of fruit. As a rule, a supernumerary bloom resulted in a low rate of fruit set. The flower or fruit set, which started growing earlier,becomes dominant in relation to other flower or fruits lagging relatively behind, this type of dominance is called primogenous.

In the flower buds of the apple, the flower of apical position is always dominant and starts growing first,as well as in gooseberry develop the first flower in basal position within the inflorescence,which is dominant than others. The dominance is due to earliness rather than their position. Those flowers are the most developed, have the best chance to growand are less exposed to be dropped.

The Role of Environmental Agents in the Drop of Flowers and Fruits

1. Climatic and meteorological conditions

Meteorological events before, during and following the development and vigour of flowers, bloom, fertilization and fruit setting are highly decisive and may influence fruit dropping. In the moderate climate, the late frosts occurring in April and May cause considerable damage on the blossom buds, flowers and young fruit primordia mainly by the destruction of important conductive tissues and cause fruit drop.

Temperatures prevailing during bloom affect the vitality and longevity of the embryo sac. High temperature than normal temperatures as well as low temperatures above the freezing point both impairs the vitality of the embryo sac. Spring frosts threaten first of all the most developed flowers within the inflorescence because the development of the flowers is related to their frost susceptibility. In apple the apical, in pear the basal flower of the inflorescence opens first,more vulnerable to frost.

Under conditions of drought, the water absorption of the leaf is stronger than that of the fruit; therefore the fruits are more exposed to be dropped than leaves. High temperature and low relative humidity caused a high degree of

June drop. In mango fruit shedding is severe when temperature is high and humidity is low. By water stress in the leaves, water was absorbed from the juicy fruits consequently they warmed up and were dropped. Young fruits lose water at 2–3- times higher rate than the ripe ones because the former are less cutinized and their surface/volume ratio is much higher.

Preharvest fruit drop is aggravated by wind and weather adversity. The pear variety,,Hardy is especially susceptible to wind. Wind with high speed promotes fruit shedding by shaking and loosening the fruits from their stalk. Red fruit drop of sweet cherry is severe, when cool and rainy weather is followed suddenly by a dry and hot period after the fruit set. Poor light condition increases the incidence of fruit drop in sweet cherry and in other fruit species.

2. *Irrigation, water supply*

Competition for the restricted amounts of water is one of the causes of fruit drop. The prevention of fruit drop immediately by irrigation is also known. The above crown water spray increases the air humidity within the canopy, which lowers the temperature of the leaves and fruits effectively by the evaporation thus reduces the danger of fruit drop. Also heavy fruit drop may be caused by excessive watering, which develops cracks in fruits like litchi, pomegranate, cherries, lemons etc.

3. *Nutrition*

Proper and balanced application of fertilizers is a pre-requisite for a tree to be able to carry its normal crop to maturity. Nitrogen being essential for a normal fruit set and in avoiding the threats of June drops. Boron plays important role in fruit set. Tree receiving high nitrogen also require high boron. Excessive nitrogen, more than 21% content in the leaves, causes a heavy drop of fruits in blueberry and accumulation of the poisonous nitrite leading to fruit drop.

However, more amount of K^+, $Ca2^+$ and carbohydrate is needed for the retention of fruits on the tree.

Biotic Factors

Pest and disease incidence on fruit plants can result in severe shedding of blossoms and fruits. Dropping of flowers in walnut is the result of *Xanthomonas* infection. Where the male inflorescencesbecome brown or black and later on, flower parts become deformed and abscise prematurely.

The fungus of fruit rot (*Moniliniafructicola*) is a significant pathogen of many temperate fruit crops like apple, pear, apricot, peach *etc.,* and is a major cause for immediate preharvest fruit drop. This pathogen is a parasite penetrates through wounds and enjoys the scares caused by hail or pests. The softened fruit is abscised at the upper end of the fruit stem without any abscission layer. The peduncle is often maintained on the tree if not abscised later fruits get shriveled and mummified within the hard pericarp during the dry weather of July.

The leaf-curl disease of peach *(Taphrinadeformans)* may also lead to fruit drop. Premature fruit drop caused by scab *(Venturiacarpophyla)* in apricot and peach is particularly dangerous during a prolonged period of drought. The first symptoms appear after the drop of petals during the 10–14th week, initially on the green fruits as pale yellowish-green spots. The fruits shrivel subsequently and are dropped. Most susceptible are the late ripening apricot and nectarine varieties.

Fruit drop of walnuts may be induced by *Gnomonia.*Gooseberry, blueberry and blackberry are haunted by grey mould *(Botrytis)* causing fruit drop. The afflicted berries display brown flecks, rot subsequently and are shed. It is largely favoured by a humid microclimate. In Citrus species, fruit drop after bloom is caused by *Colletotrichumacutatum*.

Pests

The beetle *(Omophlusproteus)* attacks sweet cherry. They destroy the pistil or young fruit primordia. The fruit used to be drilled irregularly. The hurt flower or fruit primordia is fading and drop soon. Flower parts (stamina and pistils) of fruit species (plum,apple, sour and sweet cherry, walnut and almond) are attacked by maybeetle. The hairy beetle*(Epicometishirta)*starts feeding on flowers of lower position eating preferably the pollen of the anthers, but may continue to bite the pistil too and cause fruit drop.The red backed proboscide beetle *(Coenorrhynchusaequatus)* also contributes to the fruit drop in fruit crops like apple,pear, quince, plum, almond, hawthorn and rowanberry.

The proboscide beetle of the hazel*(Balaninusnucum)* and of the oak *(Balaninusglandium)* brings up their larvae within the hard shell of the nut or acorn. The beetle is the precursor of the fungus *(Moniliafructigena)* causing fruit drop, which may attain 70%. The peach moth *(Anarsialineatella)* initiates fruit drop after having fed on the green stone fruits during the preharvest period. The attack ensues after the stone (endocarp) already hardened sufficiently. The point of penetration is either near to the fruit stem or where two fruits are touching each other. The ripening process is speeded up by the scare and a preharvest drop is initiated. On developed fruits gum used to appear. The door is also opened for *Monilia* by the caterpillars.

The moth *(Spilonotaocellana)* cause flower and fruit drop at the rate 5–10% in apple, quince, sweet cherry, peach, apricot and hazel. The scared fruits are dropped continuously being often infected by fungi and rot. The oriental fruit moth *(Grapholithamolesta)* is mainly found on peach, apricot and on raineclaude-type of plums causing fruit drop.

Fruit drop control in some important fruit crops

Mango

It exhibits maximum fruit drop among fruits *i.e.,* 90-99% which occurs in the initial 3-4 weeks after fruit set. Main causes are lack of pollination, low stigmatic receptivity,defective perfect flowers,unfavourable climatic conditions like rain,wind and cloudiness at the time of pollination,incidence of diseases like powdery mildew,anthracnose,presence of hoppers, and mealy bugs,and hormonal imbalance in fruits.Regular irrigation and proper control of pests and diseases during fruit setting and development stages decreases fruit drop to a certain extent. Growth regulator application like NAA@25-40 ppm or 2,4-D@10-15 ppm or CCC@ 200 ppm as sprays immediately after set helps to decrease fruit drop considerably. By following girdling chance of fruit drop is less.

Citrus

In all commercial varieties of sweet orange, mandarin and grape fruit,fruit drop is a serious problem.The first drop occurring in the month of May and June is mainly due to inadequate and imbalanced manuring and insufficient care.The second drop occurring in August- September is due to the attack of fungus *Colletotrichumgloeosporiodes.* The third drop is the most serious one, occurring in December and January, when the fruits are mature.This is caused by the fungus *Alternaria citri.*

Control

- Proper upkeep and maintenance of orchard.

- Optimum manuring.

- Spraying of growth regulator 2,4-D(10 ppm) alongwithaureofungin(20 ppm) and zinc phosphate(0.05%) will check physiological as well as pathological pre harvest fruit drop.

Apple

Chemicals like NAA and 2, 4, 5-T @ 20 ppm when sprayed two weeks prior to normalharvesting reduces pre harvest fruit drop in apple. Pre harvest drop in Red Delicious apple can be prevent with GA3@25 ppm, when sprayed at 5-6 week of petal fall stage. Under Chaubatia condition, 2, 4-D and 2,4,5-T@5 ppm helped in decreasing the fruit drop.

Apricot

Pre harvest drop in New Castle prevent with the help of 2, 4, 5-T, while 2, 4, 5-T@80 ppm gave the best result in Shipley Early.

Management of Fruit Crops

(a) Prompt picking

Prompt picking is an easy way to maximize the crop yield. As soon as true drop begins have the crop picked immediately. This strategy requires that growers should be aware of fruit dropping problem and regularly scout the plants for true drop. It is important that growers not jump and pick before fruit is physiologically matured or color has not reached a marketable level. "Push off" (false drop) usually occurs several days before true physiological drop and premature picking can result in poorer fruit quality and storage.

(b) Thinning

To avoid fruit drop as a result of overbearing, thinning can be recommended for the young fruit before the tree drops it. In general, it is best to leave 4-6 inches between each fruit and break up any clusters that may form. For good fruit size as well as avoiding biennial bearing, apples and pears should be thinned so that there is six to eight inches between fruits, with four to six inches between peaches, and two to three inches between plums. Apricots should be far enough apart so fruits do not touch. Small, sharp pruners may be used to remove the fruit or simply pluck it off with your hands. If you pinch the blossoms off your tree before the petals drop and fruit begins to form, you will also be able to help prevent fruit drop.

(c) Water and moisture management

Moisture stress during fruit set and fruit development causes severe fruit drop in most of the fruit crops. Too much water can also lead to fruit drop.Hence proper water management is required. In the deep rooted perennial fruit trees, the concept of irrigation scheduling based on plant water content should be ideal one.Conservation of soil moisture by mulching enhances water use efficiency. Adoption of drip irrigation system ensures adequate amount of water supply to the crop throughout the growth period. Irrigation is with held when ripening starts to facilitate harvesting, hasten fruit ripening and to reduce fruit drop caused by high humidity.

(d) Temperature

Mango trees do best in warm climates such as those found in subtropical and tropical locations, on the other hand they are very much prone to frost injury.If the temperature drops to

40 F or lower, small fruit and flowers may drop off the branches after a few hours. Protect the tree from frost during its first two years by providing an overhead covering, applying water or heat, or increasing air circulation.Another option is to help protect the trunk by wrapping it with straw or foam tree wraps. Do not prune dead branches or twigs until the danger of frost has passed.

(e) Nutrition management

A balanced fertilizer programme should be adopted to control fruit drop.Before application of fertilizer, soil testing is necessary to confirm deficiency or toxicity of nutrients in soil. Integrated nutrient management concept may be the most suitable one for supply of optimum level of nutrients to fruit crops

(f) Pest and disease management

Suitable insect control strategy should be initiated to minimize their harmful effects. Be aware that using pesticide sprays while your trees are in bloom may terminate bees and other beneficial insects, and some sprays may even cause fruit drop.

(g) Pollenizers

These are the cultivars which provide abundant viable and compatible pollengrains to the main crop. Adequate number of pollinizing cultivar(s) should be planted at the time of planting. In case of apple, plant 2-4 rows of a particular variety and alternate with 2-4 rows of another variety with the same or an overlapping blooming period. Breeding and selection of self fertile cultivars may reduce difficulty of achieving satisfactory pollination.

The pollinizer variety should not be farther away. It is always desirable that more than one pollinizing variety may be selected for planting an orchard with 33 per cent pollinizing trees.It is very convenient to plant one row of pollinizing variety after every two rows of the commercial variety. With more than one pollinizing variety, it is recommended to plant them in alternate combinations of the standard and pollinizing cultivars.

(h) Planting windbreaks and shelterbelts

High wind velocity causes mechanical damage to the fruits and branches and also their desiccation due to excessive transpiration loss of water.Suitable windbreaks and shelterbelts are to be grown to provide a protective barrier against hot and cold wind and to reduce the extent of fruit drop.

(i) Use of growth regulators

In different fruit crops, different growth regulators are recommended to control fruit drop.Recommended dose should be strictly followed, for e.g.2, 4-D at lower dose acts as a hormone, while at higher dose it acts as a herbicide.In case of mango NAA or 2, 4-D spray @20- 30 ppm during last week of April, when the fruits attain marble size, effectively controls fruit drop. Incitrus, two sprays of 2, 4-D or GA3 @8-10 ppm and 50 ppm before the young fruits attain growth, checks fruit drop considerably.

ReTain (aminoethoxyvinylglycine or AVG) - This material has proven to be more effective than NAA for drop control but has the drawbacks of being harder to time application and more expensive to use.ReTain, may increase soluble solids, color, fruit size, fruit firmness and reduce the increase of water core. It delays harvest 7-10 days and must be picked when maturity indices indicate. The recommended rate is 17oz (of 333 grams)/100 gallons.

References

Berry, F. H., 1960. Etiology and control of walnut anthracnose. Univ. Md. Agric. *Exp. Stn. Bull.* A-113: 22.

Catlin, P. B., Ramos, D. E., Sibbett, G.S., Olson, W. H. & Olsson, E. A., 1987. Pistillate flower abscission of the Persian walnut. *Hort. Science* 22 (2): 201–205.

Gomez-Cadenas, A. – Mehouachi, J. – Tadeo, F.R. – Primo-Millo, E. – Talon, M., 2000. Hormonal regulation of fruitlet abscission induced by carbohydrate shortage in citrus. *Planta.* 210:636–643.

Goren, R., 1993. Anatomical, physiological, and hormonal aspects of abscission in citrus. *Hort. Rev.* 15: 145–182.

Greene, D.W., 1989. CPPU influences McIntosh apple crop load and fruit characteristics. *Hort. Science.* 24: 94–96.

Holb, I. J., 2004a. The brown rot fungi of fruit crops (Monilinia spp.) II. Important features of their epidemiology (Review). *International Journal of Horticultural Science*, 10 (1): 17– 35.

10

Harvesting and Packing of Fruit Crops

Madhubala Thakre

Division of Fruits and Horticultural Technology, ICAR-IARI, New Delhi

After doing all hard work during fruit production the last operation is harvesting on field. Though it is last operation, it should be done very carefully to get the quality harvest. Harvesting of fruits mainly depends upon the nature of fruit. Here nature refers to the classes of fruits divided on the basis of their respiratory pattern and ethylene production. On the basis of the respiratory pattern the fruits can be divided in to two categories, first one is climacteric and the second one is non climacteric.

(1) **Climacteric fruits:** The meaning of word „climacteric is a critical period or event. Here, climacteric refers to the event of respiratory rise before the ripening phase. Those fruits which show such patterns are known as climacteric fruits (Taiz and Zeiger, 2002) Examples: Mango, guava, apple, banana, sapota, papaya, peach, pear, plum, fig, *Annona.*

(2) **Non climacteric fruits:** Non climacteric fruits do not exhibit the respiration and ethylene production rise (Taiz and Zeiger, 2002). Examples: Citrus, grapes, pineapple, pomegranate, litchi, ber, jamun, cashew, cherry, strawberry.

Climacteric fruits can be harvested after the maturity and can be ripe during storage. Whereas, the non climacteric fruits harvested after ripening on the tree. If they are harvested before proper ripening then they won t get the appropriate quality. Each fruit has its own maturity indices (Table 1).

Table 1: Maturity indices and harvesting method varies according to fruit crop (Chadha, 2001).

S.No.	Fruit	Maturity indices	Harvesting Method
1.	Apple	TSS, ease in separation of fruit from spur, change in ground surface colour from the green to pale change in the seed colour to light brown, fruit firmness days from full bloom to harvest	Hand picking
2.	Apricots	TSS	Limb shaking
3.	Bael	Mature green fruit	Individual fruits along withstalk

Contd.

4.	Banana	-Number of days from flower emergence - Disappearance of the angles - Pulp:peel ratio -Brittleness of the floral remnants and their natural shedding -Dullness of the fruit skin colour	Cutting by knife
5.	Ber	Skin colour, Days from flowering to maturity	Hand picking
6.	Cherry	Colour development, TSS and flavor	Hand picking
7.	Fig	-Soft slightly wilted at the neck and droop andlittle or no milky latex flow at the cut end of the stalk	Hand picking
8.	Grapes	Seeded varieties: Dark brown seeds Seedless varieties: Berry colour	Hand picking
9.	Guava	Skin colour	Individual fruits are harvested by hand
10.	Jamun	Skin colour	Hand picking, shaking of branches and collecting fruits in a polythene sheets
11.	Kiwi fruit	TSS 6.2%	Hand picking
12.	Limes and lemons	Colour break stage (mature but still green)	Hand Picking
13.	Litchi	TSS, Days from fruit set	Litchi is harvested by picking fruits along with some leaves with the help of sharp instrument.
14.	Mandarin	Peel colour, TSS: acid ratio	Hand picking
15.	Mango	*'Tapka'*, specific gravity, Days from flowering	Hand picking, using mango harvester
16.	Olive	Skin colour	Hand picking
17.	Papaya	Skin colour, latex become watery	Hand picking
18.	Peach	Days to maturity, calendar date, fruit size, firmness, sense of touch, pit discoloration, freeness of pit, taste, ground colour, sugar: acidity ratio.	Hand picking
19.	Pear	Days to maturity	Hand picking
20.	Phalsa	Fruit colour	Hand picking
21.	Pineapple	15-18 months after planting	Cutting with knife
22.	Plum	Days from full bloom, firmness, TSS, change ofground colour from green to yellow of red depending upon cultivar.	Hand picking
23.	Pomegranate	Closing of calyx, grain cracking sound when pressing from outside.	Hand picking
24.	Sapota	At maturity no green tissue and milky latex isvisible	Hand picking
25.	Strawberry	Half to one-fourth of skin develops colour	Hand picking
26.	Sweetorange	Peel colour, TSS: acid ratio	Hand picking
27.	Walnut	PTB (packing tissues turns brown) stage	Knocking of trees after spreading polythene sheets on ground beneath trees

Pre-cooling: Pre-cooling is the practice which aims to remove field heat of freshly harvested produce. It reduces the deterioration and slows down the metabolism. Pre-cooling minimizes the losses by rapidly lowering the temperature. In some fruit crops it also reduces disorders like chilling injury. The simplest way of pre -cooling of fruits is to keep them in shade. Other method of pre-cooling includes air, hydro and vacuum cooling.

1. **Air cooling:** It is done by placing the fruits in the cold room in well ventilated containers.

2. **Hydro cooling:** In this method of pre-cooling the fruits the dipped or sprayed with cold water.

3. **Vacuum cooling:** It is the costliest method of pre-cooling. In this method of pre-cooling the atmospheric pressure is reduced which results in the evaporation of water from fruits. This reduces the temperature of fruit.

Grading: It is the operation in which fruits are divided in the different grades as per their size or colour or any other parameter specific to the fruit. This helps the grower to get better prize as well as the consumer to get quality product with respect to price of the commodity. Grades vary according to fruits.

Packaging and packing: Packaging, the technology used to contain, protect and preserve products throughout their distribution, storage and handling and at the same time to identify them, provide instructions for their use and promote them."Packing, on the other hand, is the process of placing the items that have already been packaged into larger containers for shipping. Packaging is very important before packing. In general fruits are packaged in CFB boxes. These boxes are available in different sizes along with different degree of ventilation. Some information like name of fruit and variety, quantity in box is mentioned over boxes. Different kinds of cushioning materials such as paper cuttings, fruit foam covers can be used during packaging.Some other types of packaging are also available like modified atmospheric packaging and vacuum packaging.

1. **Modified atmospheric packaging:** In this type of packaging a fixed gas mixture is introduced which replaces the existing air of the package. There will be no further control over the composition of air inside the package.

2. **Vacuum packaging:** In this type of packaging, the air is removed from the package and fruit is covered with polythene sheets. In this packaging, firstly the product is packaged in a air tight container and then air removed and then sealed. Lack of oxygen helps to reduce respiration, oxidation and development of oxygen-breathing microorganisms.

After packaging, fruits are packed in comparatively large containers made of wood or corrugated fibre boxes depending upon the weight of the fruits.

Transport: Depending upon the distance, fruits can be transported by roads, rails, air freight or marine containers. Transportation to nearer markets can be done by road ways. In such type of transportation the CFB boxes are loaded providing some cushioning of straw at the bottom of trolley. Sometimes this type of transportation is very faulty by placing fruits directly on the cushion without any prior packaging. It results in brushing of fruits and losses due to injury like rotting. Now a day refrigerated vehicles are available for transportation of some perishable fruits like litchi. They can be used for safe transportation via roads. Distant transportation within country can be done by railways. But, the foremost requirement is the proper packing and packaging of fruits to resist the rough handling during loading and unloading. Transportation with air freight or marine containersis costly and usually practices during import and export. There are some factors which affects fruits during transportation. These include initial quality of produce, temperature, humidity and water loss, atmospheric composition, mixed loads and physical injury (Vigneault *et al.*, 2009).

1. **Initial quality of produce:** It refers to the maturity at harvesting and freeness to other damages. Fruits are harvested at comparatively more mature stage for local market selling as compared to distant market selling. The produce should be free from any mechanical damages and any other factors affecting their quality during transportation.

2. **Temperature:** Respiration is the mechanism by which the fresh fruits remain alive. Temperature is the most important factor that affects the fruit quality during transport by affecting respiration. So, the commodity should hold at its lowest recommended storage temperature.

3. **Humidity and water loss:** In general fresh fruits require 90-95 % relative humidity during transportation. But in most cases, it is not maintained up to the required one. Hence, affected the produce quality by water loss. However, marine containers have better provisions to maintain humidity as compared to the aeroplane. But, temperature should be controlled along with humidity to stop ice formation and chilling injury to the produce.

4. **Atmospheric composition:** Containers used during highway (road) transportation have sufficient ventilation to exhaust the CO_2. But, marine containers are very tight. So, the CO_2 generated due to respiration should be exhausted by enabling them with efficient controlled atmospheres systems.

5. **Mixed loads:** Different types of fruits should be load at different carriers.So that,their transportation requirement can be fulfilled properly. Because it is possible that some conditions that are favourable for one fruit is not favourable for other.

6. **Physical injury:** If produce is not properly packed, then there are chances of several types of physical injury due to compression, rubbing etc. So, appropriate packaging along with proper cushioning is required to protect fruits from physical injury.

References

Chadha, K.L. 2002. Handbook of horticulture.Indian Council of Agricultural Research.107-338.
    Taiz, L. and E. Zeiger. 2002. Plant Phyisiology. Third edition. Panima Publishing Corporation, New Delhi. pp.524.
Vigneault, C., Thompson, J., Wu, W., Hui, K. P. C and D. I. LeBlanc. 2009. Transportation of fresh horticultural produce. *Post harvest Technologies for Horticultural Crops*. 2: 1-24

Colour Plate Section of Part-I: Theory

1: Orchard Layout and Establishment of Fruit Orchard – Principles and Practices

Fig. 1: High density orchard of mango variety Amrapali

Fig. 2: High density plantingin Papaya 6,400 plants/ha (1.25 x1.25 m)

Fig. 3: High density planting in Guava

Fig. 4: High density planting of Kinnow mandarin on troyer citrange rootstock [Planting density = 3,086 plants/ ha (1.8 m x 1.8 m)]

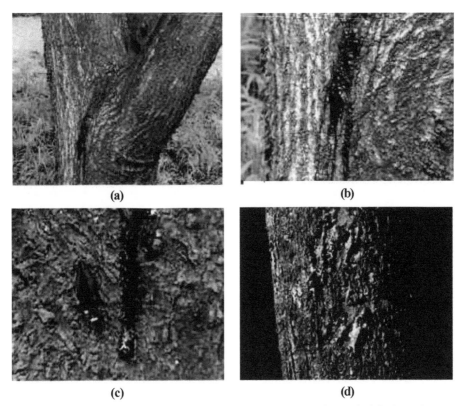

(a)   (b)

(c)   (d)

Fig. 5: Mango decline tree showing gum secretion from **(a)** main trunk right from the tree base, **(b)** middle portion of tree, **(c)** tree top and **(d)** side branches before planting turmeric as intercrop

Fig. 6: Larvae inside the trunk of a declining mango tree

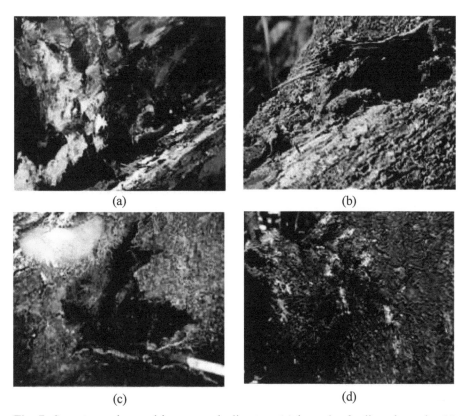

Fig. 7: Symptoms observed in mango decline tree **(a)** irregular feeding channels, **(b)** webs **(c)** vascular discoloration and **(d)** outer wood showing cracks

Fig. 8: Turmeric plantation as intercrop in mango declining orchard

Fig. 9: Mango trunk showing no symptoms of gummosis after planting turmeric as intercrop in declining mango orchard

Fig. 10: Mango decline tree showing symptoms of dead branches and dry scorched leaves at tree top

Fig. 11: Mango decline tree from control block showing **(a)** Irregular feeding channels in transverse section of trunk, **(b)** feeding channel and vascular discoloration in transverse section of trunk, **(c)** trunk borer below the bark, **(d)** vascular discoloration below the gummosis region.

2: Frost, Frost Protection and Winter Injury

Fig. 1: Banana leaves damaged due to frost injury

Fig 2: Severe frost injured citrus tree   **Fig. 3:** Frost injured mango branches

Fig. 4: Olive tree injured with frost   **Fig. 5:** Papaya frost injury

Fig. 6: Strawberry fruits affected
with frost injury

3: Bearing Habit of Fruit Crops

Fig. 1: Mango panicles
(Bearing habit: Group-I)

Fig. 2: Apple
flowers (Bearing
habit: Group-II)

Fig. 3: Guava flowers
(Bearing habit: Group-III)

Fig. 4: Kinnow sprouting and flowering (Bearing habit: Group-IV)

Fig. 5: Litchi flowering (Bearing habit:
Group-V)

Fig. 6: Baser flowering (Group-VI)

Part-II: Practicals

Anil Kumar Dubey, Om Praksh Awasthi, Sanjay Kumar Singh and V.B. Patel

1

Identification of Fruit Varieties

1. Mango

Varieties released from I.A.R.I., New Delhi

Malliaka: (Neelum x Dashehari)

This variety was released in 1971. Tree moderate, semi-vigorous to vigorous, moderate spreading, bearing moderately and regular. Fruit large, ovate-oblong; beak slight, sinus absent, peel thick, apricot yellow; flesh firm, fibreless. Fruit quality superb with excellent sugar acid blend., TSS (24-20° Brix), good keeping quality (15 days). It is suitable for table and canning purposes. Maturity during $3^{rd} - 4^{th}$ week of July.

Amrapali: (Dashehari x Neelum)

This cultivar was released in 1979. Tree small, dwarf and medium spreading and highly suitable for high density plantation. Fruit small to medium (130 g), ovate-oblong; beak indistinct, sinus slight, peel thick, light greenish apricot yellow, somewhat fibrous; flesh firm, deep orange red, fibreless. Fruit quality excellent. Maturity in 4th week of July.

Pusa Arunima: (Amrapali x Sensation)

This mango hybrid was released in 2002. Regular bearer, semi-vigorous and suitable for closer planting (6 m x 6 m; ripening 4th week of June to 1st week July in north India. Fruits medium to large (230 to 250 g), attractive red peel; medium TSS (19.5° Brix), suitable for both domestic and international markets. It has long shelf-life (10 to 12 days) at room temperature after ripening.

Pusa Surya: (Selection from Eldon)

Released in 2002 as a selection from an exotic variety Eldon introduced from Brazil. Trees are semi-dwarf and suited for close planting (6 m x 6 m). Fruit medium to large (260 to 290 g), attractive apricot yellow peel; pulp medium, TSS (19° Brix) with long shelf-life (10 to 12 days) at room temperature after ripening. Fruit ripens by mid-July in north India,

Pusa Shreshth (Amrapali x Sensation)

Plant medium, semi dwarf, regular and moderate tolerance to major pests. Fruit medium (211-241g), attractive, elongated shape; peel red, pulp orange, pulp content, 71.9 g. pulp: stone ratio, 3.7-4.6, TSS, 18.5-21.5o Brix, acidity 0.15-0.25%, ascorbic acid content, 38.80-40.5 mg/100g pulp.

Pusa Pratibha (Amrapali x Sensation)

Plant is medium statured with lanceolate leaves; semi-vigorous growth habit; bearing regular, Fruit medium (181 g), attractive, elongated oblong; peel bright red, pulp orange, high pulp content (71.1%), good TSS and acid blend. TSS 19.6%, ascorbic acid (34.89 mg/100 g pulp), shelf life 7 to 8 days at room temperature; mature after 140 days of flowering.

Pusa Peetamber (Amrapali x Lal Sundari)

Plant short statured with lanceolate leaves, semi-vigorous; suitable for planting at a distance of 6 m x 6 m; Fruit medium (213.0 g), elongated oblong, yellow; good TSS and acid blend; TSS 18.8%; shelf life 5 to 6 days at room temperature; high pulp content (73.6%) and ascorbic acid content (39.78 mg/100 g pulp). Mature after 140 days of flowering;

Pusa Lalima (Dushehari x Sensation)

Plants are; medium statured plant with lanceolate leaves, semi-vigorous growth habit, suitable for planting at a distance of 6 m x 6 m. Fruit medium (209.0 g), attractive, oblong, bright, peel red, pulp orange. TSS 19.7%, high pulp content (70.1%) and ascorbic acid content (34.73 mg/100 g pulp), shelf life 5 to 6 days, mature 125 days after flowering

Varieties Released from IIHR, Bangalore

Arka Anmol: (Alphanso x Janardan Pasand)

Tree medium vigourous, regular, prolific bearer, fruit weight is 250 g with uniform, peel yellow, pulp orange, TSS 20o Brix, keeping quality is excellent, this cultivar is free from spongy tissue disorder and suitable for export.

Arka Aruna (Banganpalli x Alphonso)

Tree dwarf, regular and precocious bearer; it is late maturing cultivar. Fruit large (500-750 g), pulp high (73.378.5%) and free from spongy tissue, TSS 22° Brix.

Arka Neelkiran (Alphonso x Neelum)

Semi-vigorous, regular bearer. Average weight of fruit 340 g, pulp content is 68%, pulp orange, TSS 210 Brix. This cultivar is also free from the spongy tissue disorder.

Arka Puneet (Alophonso x Banganpalli)

Tree vigorous, regular,prolific bearer. This is a "mid season cultivar. Average fruit weight is 284 g, pulp content is 75%, orange yellow. Free from fibre. TSS 21° Brix, the flavour of fruit is just like Alphonso. This cultivar is also free from the spongy tissue and fruit fly.

Arka udaya (Amrapalli x Arka anmol)

Late-season variety, sweet in taste, high-yielding with a long shelf-life. Fruits can stay fresh at room temperature for about 10 days without refrigeration. Regular bearer.

Varieties Released from CISH, Lucknow

Dashehari-51

A regular bearing and high yielding clone of Dashehari has been selected from the orchard of CISH, Lucknow and released during 1998. Per year productivity of this clone is 38.8 per cent more than that of the normal Dashehari even in off year. This clone produces good crop every year without 'off bearing rhythm'.

CISH-M2 (Dashehari x Chausa)

Fruits are dark yellow in colour with firm flesh and scanty fibre. It is a late season variety and has good commercial value.

Ambika (Amrapali x Janardan Pasand)

Fruits yellow with dark red blush and firm dark yellow flesh. Fruit medium (225 g), oblong in; sinus slight, beak broadly pointed. Peel smooth and tough. TSS 210 Brix. It is late in ripening hybrid.

Varieties released from RFRS, VEGURLA Ratna (Neelum x Alphonso)

This variety was released in 1981. Tree very dwarf in North India with weak and spreading branches. Newly emerged leaves are dark red and small to medium; regular bearing. The fruit maturity is late (fruit mature in 3rd week of July) in North India. Fruits medium to large (315 g), peel deep yellow, pulp recovery 79% and orange. The fruit is free from spongy tissue. The TSS is 23.8° Brix, acidity 0.25% and fruit taste is sweet.

Sindhu (Ratna x Alphonso)

First seedless variety released in 1992. Tree semi-vigorous, regular bearer, fruit medium (215 g), attractive; peel red, pulp deep orange, pulp to stone ratio high (26:1), fibreless, free from spongy tissue. Taste of fruit is pleasant with a better sugar acid ratio than Ratna. TSS above 22° Brix and acidity medium (0.25%). The stone of fruit is very.

Konkan Ruchi (Neelum x Alphonso)

Released in 1999, regular bearing mango variety developed for pickle making. The tree is heavy yielder, fruit large (430 g) and pulp recovery is about 78%.

Other Varieties

Alphanso

Tree medium, bearing low and biennial. Leaves oval- lanceolate, out- held, twisted; apex in moderate curve, sinus absent, peel medium to thick; fruit quality good, flavor pleasant, taste very sweet, keeping quality good; stone medium. Moderately resistant to wind and hopper.

Kesar

It is a commercial variety of Gujrat and second to the Alphanso in terms of export. Fruits medium to large (3-4 fruits/kg), taste is very good, sugar acid blend excellent. It is moderate yielder and biennial bearer.

Dahsehari

It is a midseason and heavy bearing cultivar. The tree medium, moderate vigorous, spreading; rounded, medium to dense canopy. The fruit primrose to canary yellow with abundant light yellow dots, medium; peel smooth. The flesh yellow, firm, with almost no fibre and a delightful aroma and very sweet taste.

Langra

Tree tall, and spreading, bearing medium and biennial. Fruit oblongish, beak much prominent, sinus not marked, peel medium, lime green; flesh firm, lemon yellow, fibre scanty near peel. Fruit quality very good, pleasant aroma.

Samarbehisht Chowsa

Originated as superior chance seedlings in Chowsa village, Lucknow. The tree moderately vigorous and spreading. Fruit medium, ovate to oval-oblique, beak distrinct to prominent, sinus slight, peel medium thick, flesh firm, umber yellow; fibreless. Pulp soft and juicy with scanty fine, long fibre near the skin.

Bombay Green

The tree large and speading. The fruit medium, ovate-oblong to oblong-reniform, beak absent, sinus shallow, peel thick, spinach green; flesh firm to soft with scanty fibre just under the peel, very sweet with pleasant aroma.

Himsagar

The tree vigorous, tall, dense, spreading; the fruit large (450-550 g), greenish yellow to bright yellow with no blush, with light yellow dots, ovate with flattened base. The peel thin, tough and easily separated; flesh firm and juicy with no fibre, orange, rich and sweet with a mild aroma, of good to excellent quality.

Banganpalli

The tree medium, moderately vigorous and spreading, bearing heavy and regular. Fruit large, obliquely oval, sinus shallow, peel thin, golden, flesh firm to meaty, fruit quality good, juice moderately to abundant without fibre.

Kurakan

Tree medium, top rounded, bearing heavy, fruit medium to small, oval. Beak absent to a point; sinus slight, peel medium thick. Flavor aromatic, taste medium sweet. It is a polyembryonic cultivar.

Pusa Shreshth   Pusa Arunima   Pusa Pratibha   Pusa Arunima

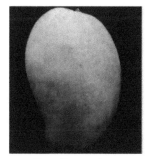

Pusa Peetamber   Pusa Lalima

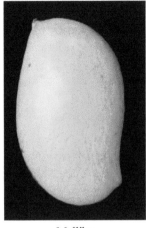

Mallika

Ambika

2. Citrus Varieties and Species

I. M andarin (*C. reticulata*)

The mandarin group comprised numerous species as well as intergeneric and interspecific hybrids, which possess several unique characteristics. The most distinguishable trait of mandarin group is easy peeling character. Mandarihs are divided into five groups:

i)   Mediterranean mandarins (C. *dcliciosa* Tan.)

ii) Satsuma (C *unshiu* Marc.)

iii) King mandarin (C *nobilis* Lour.)

iv) Common mandarins (C *reticulata* Blanco)

v)   Small fruited mandarins.

A widely grown mandarin is Ponkan which is of great significance in India, China, the Philippines and Brazil. The detail description of mandarin cultivars is mention here under.

Khasi Mandarin

Tree is generally medium to tall with erect habit, densely foliage, fruits depressed, globose to oblate, orange yellow to bright orange, surface smooth, glossy, base even, short necked, rind thin soft, adherence very slight, juice abundant, orange coloured.

Nagpur Mandarin

Large sized tall tree, trunk robust, branches growing up right, foliage not dense, bark thin, brown, spines absent. Petiole cylindrical, long, articulate, petiole wings narrow, upper surface dark green, smooth, glossy, aroma distinct. Flowers mostly solitary in the axil of leaves, flowers bisexual, bud obovate, star shaped at the apex, pedicel short. Fruit depressed, globose to oblate, colour yellow, surface smooth, glossy, stem short, strongly attached. Rind thin, soft to very soft, oil glands numerous, juice abundant, TSS 6.5%, acidity 0.735%, flavour good, taste good blending of sour and sweet. Seeds 11/fruits.

King (*C. nobilis*)

Tree stiff and upright growth, generally thorny, frequently hanging branches, leaves small dark green, petiole narrowly winged. Fruits large and mandarin like with rough and bumpy, deep orange yellow, base rounded, easily separable, juice abundant, quality very good.

Dancy (*C. tangirina*)

Best known of the Mandarins. Tree has fine foliage and upright habit. Good flavor. Plant medium sized, erect growing, tall tree, crown compact, rounded, trunk robust, branches growing up right, bark thin, brown, angular, spines absent, petiole cylindrical, articulated, medium long, narrow winged, broadly lanceolate, size small, upper surface dark green, glossy smooth, apex, flower small, bisexual, white in colour, rarely male flowers seen at the end of flowering season, pedicel short, tampering towards the base, pollen abundant. Fruit easy to peel, oblate to pyriform, colour reddish orange, surface smooth, finely pitted, tips blunt, stem short, rind thin, soft leathery, oil glands average, aroma distinct spicy, juice moderate, TSS 6.5% and acidity 0.78% , some seeds.

Willow leaf (*C. deliciosa*)

A large to medium tree, almost thorn less, spreading, broad topped, very willowy in growth (drooping branches). Leaves small, narrow, deep green, sharp pointed. Fruits small to medium with thin rind, strongly compressed, rind and segments loose, juice plenty, juice sacs short.

Satsuma (*C. unshui*)

Hardiest of all mandarins. Seedless and easy to peel. Excellent flavor when ripe in winter months. A very slow grower. Tree thornless and spreading habit, round topped, leaves broad, tapering abruptly towards the apex, petioles scarcely margined, flower medium in size, fruits medium to small, compressed, orange at maturity and pulp orange. Very juicy, acidity and sweetness well blend, easily separable. Polyembryonic species.

Kara Mandarin

Tree moderately vigorous, spreading and round topped, similar to Satsuma but larger and more vigorous, thorneless with rather stout, spreading and drooping branches, leaves dark green, hardy to cold. Fruit medium large, moderately to slight oblate, base commonly slight necked and furrowed, apex flattened or depressed with visible areolar area. Rind medium thick, soft to texture, moderately adherent but peels freely well, surface slightly rough and bumpy.

II. Sweet Orange (*C. sinensis*)

Sweet orange is the most widely distributed and exhibits the highest production among all commercial citrus species. Sweet oranges can be divided into four groups based on fruit norphological characteristics, chemical constituents and usage.

i) Common or round oranges: These include Valencia, used for fresh fruit and processing, Shamouti, with its typicai form and flavour, and Pera, Hamlin and Pineapple, grown mainly for processing.

ii) Navel orange: This is the most important group for fresh fruit. Navel orange has the prominent distinctive feature of a small, secondary fruit embedded in the apex of the main fruit.

iii) Pigmented (blood) orange: Anthocyanin appears in the rind and juice. The best known varieties are Moro, Tarocco and Sanguinelli.

iv) Acidless or sugar orange: This has very low acidity in the fruit (about 0.2%). Mosambi, Succary.

Following are the cultivars of sweet orange.

Satgudi

Fruit almost spherical, small to large, smooth surface, attractive orange coloured when fully matured, base and apex is fully rounded, rind thin with little rag, semi glossy and finelly pitted, pulp uniformly straw colour, juicy, flavour excellent, seeds few to many.

Italian.

Washington Navel orange

Famous winter-ripening fruit. Excellent in flavour. Seedless, easy to peel, separates into segments. Also good for fresh squeezed juice in winter.

Valencia Orange

The traditional juice orange, also good for eating. Blooms in spring and has small green fruit first summer which ripen the following summer. Stores well on tree for long periods actually improving in quality.

Jaffa Orange

Also known as 'Shamouti' orange. Fruit is almost seedless, pleasantly sweet and juicy. Easy to peel. Stores well on tree.

III. Lime (*C. aurantifolia*)

Lime trees most probably originated in tropical areas along the Malay archipelago and as a result of this heritage are the most freeze sensitive of all commercial citrus species. Thus, the distribution of limes is limited to the tropics and warm, humid subtropical regions where minimum temperatures remain above -2° C to -3°C. Acid limes consist of small fruited Indian, West Indian or Mexican lime (C *aurantifolia)* and large fruited Tahiti or Persian lime (C *latifolia).*

Tree medium sized, hardy and semi vigorous, upright with an irregular and loose crown. Foliage not dense, light green, thorns numerous, leaves broadly elliptical, margin crenate, apex obtuse, veins prominent, petiole narrowly winged, distinctly articulated, fruit round to oblong, yellow, smooth, apex rounded and slightly nippled, base rounded, rind thin, papery, adherence very strong, segments 9-11, pulp light greenish-yellow, juicy, flavour good and sour.

Bears Seedless Lime

Large fruit, very juicy and seedless. Most be protected from severe frosts. Crop heaviest in fall, although fruit stores on tree finally turning from green to yellow before falling off

Mexican Lime

Known as 'Key Lime in Florida, this is the most tender citrus. Fruit much smaller than 'Bearss' with many seeds. Strong lime flavor and Juicy.

Tahiti Lime

'Tahiti' tree are monoembryonic, indicating that this variety is not completely female sterile. The progeny were variable, resembling lemon and citron.

Kagzi Lime

A small tree with rather irregular branches; twigs with short, stiff, very sharp spines; leaves small, 5-7.5 cm long, elliptic-ovate or oblong-ovate, obtusely pointed at the tip and rounded at the base, margins crenulate, pale green, petioles narrowly winged, spathulate; inflorescences axillary, short, lax racemes of 2-7 flowers (rarely single); flowers small, white in the bud, calyx cupulate, 4-5 lobed; petals 4-5, stamens 20-25; ovary depressed, globose, with 9-12 segments, not merging into the style but clearly set off from it; style soon deciduous; stigma depressed, globose; fruits small, oval or subglobose, often with a small apical papilla, greenish-yellow when ripe; peel very thin, prominently glandular-dotted; seeds small, oval, white inside. Pramilini, Vikram, Jaidevi, Saisarbati, ALH-77 are some important varieties of Kagzi lime.

IV. Lemon (*C. limon*)

The true home of the lemon is unknown, though some have linked it to northwestern India. Lemon trees probably originated in the_Eastern Himalayan regions of India. It is generally accepted that lemon is a hybrid closely related to citron. Distribution and major production areas are limited to semi-arid to arid subtropical regions having the minimum temperatures greater than - 4°C. Certain lemons may be hybrids between Lemon and citron. Another lemon Meyer (C. *meyerii)* is less acidic and rather resembles an orange in shape. Following are the cultivars of lemon.

Assam Lemon

It is an indigenous variety of Assam and grows all over south India, with the name Pat nimboo in Bombay-Deccan region and as Seville lemon in Andhra Pradesh. The cultivation of this variety is gaining populality because of heavy fruiting, the big size of the fruit and the abundant juice content. It is known as Sivakasi lemon or Nepali Oblong.

Bearss

Closely resembles 'Lisbon'. It is highly susceptible to scab and greasy spot and oil spotting.

The tree is vigorous and tends to produce too many water sprouts. The peel is rich in oil.

Eureka

It originated from seed taken from an Italian lemon (probably the 'Lunario'). The fruit is elliptic to oblong or rarely obovate, with moderately protruding nipple at apex, a low collar at the base; peel yellow, longitudinally ridged, slightly rough because of sunken oil glands, medium-thick, tightly clinging; pulp greenish-yellow, in about 10 segments, fine-grained, tender, juicy, very acid. Fruits are often borne in large terminal clusters unprotected by the foliage. Trees are of medium size, almost thornless, early-bearing, prolific; not especially vigorous, cold-sensitive, relatively short-lived.

Kagzi Kalan

It is originated from chance seedling. Trees are spreading, less thorny, smooth branch, leaves big. Fruits are often borne in large terminal clusters, big size, less acidic, seedless, and juicy. Bear flowers in two seasons.

Lisbon

Originated in Portuga possibly as a selection of 'Gallego. Fruit almost identical to Eureka, elliptical to oblong, prominently nippled at apex, base faintly necked; peel yellow, barely rough, faintly pitted, sometimes slightly ribbed, medium-thick, tightly clinging; pulp pale greenish- yellow, in about 10 segments, fine-grained, tender, juicy, very acid, with few or no seeds. Fruit is borne inside the canopy, sheltered from extremes of heat and cold. Trees are large, vigorous, thorny, prolific, and resistant to cold, heat, and wind. It is low-yielding and short-lived in India.

Meyer

It is a hybrid, possibly lemon x mandarin orange. Tree small, with few thorns, prolific, cold-resistant; produces few water sprouts. Plant has a mounding habit, nearly thornless. Fruits abundantly in winter but can have some fruit most months of the year. Mature fruit turns from yellow to slightly orange in color. Obovate, elliptical or oblong, round at the base, occasionally faintly necked and furrowed or lobed; apex rounded or with short nipple of medium size, peel light-orange with numerous small oil glands, pulp pale orange-yellow, usually in 10 segments with tender walls, melting, juicy, moderately acid with medium lemon flavor; seeds small, 8 to 12. It is a carrier of a virulent strain of tristeza.

Nepali Round

It originated in India; round, without distinct nipple; juicy; seedless. Tree large, vigorous, compact, nearly thornless, medium-prolific. Successfully cultivated in South India.

Villafranca

It is believed to have originated in Sicily; closely resembles 'Eureka'; of medium size. Tree is more vigorous, larger, more densely foliaged, and more thorny than 'Eureka' but becomes thornless with age.

V. Grape fruit (*C. paradisi*)

Grapefruit is probably a natural hybrid of pummelo as a seed parent and sweet orange or some other similar type as pollen donor. Grapefruit is one of the few citrus types that originated in the New World, probably in the West Indies. The high heat requirement limits production of the highest quality fruit in tropical and hot, humid subtropical regions. It has achieved prominence in the twentieth century as fresh fruit and for processing. It is not clear whether the name was given because the flavour resembling the grape or, more probably, because the fruits are borne in clusters, contrasting with the single fruit in pummelo. Following are the cultivars of the grapefruit.

Foster

Plant fairly vigorous, medium sized tree, crown rounded, compact, trunk medium, robust, branches spreading, young shoots light green, minutely hairy, short slender spines are in one or two yeas old shoots. Petiole medium long, rounded, articulated, broadly winged, shape ovate to subcordate, tip touches the very broadly rounded base of the lamina, lamina medium large, upper surface dark green, glossy, slightly leathery, apex rounded, base broadly rounded, aroma mild. Inflorescence in cymose cluster, bud ovate, oblong ovate, white, minutely hairy at apex, fruit globose to oblate, colour yellow with a blush of pink, surface smooth, coarsely pitted, base rounded or slightly depressed, stem medium long, apex rounded, areole absent, rind medium thick, firm, leathery, juice abundant, mesocarp thick, yellowish with pinkish, TSS 7%, acidity 0.87% seeds average 34 per fruit.

Duncan

Plant medium large, crown compact, rounded, trunk medium, robust, bark rather thick, gray, branches thick, spreading, short slender spines on one or two year old shoots in the axils of leaves. Petiole articulate, long, cylindrical in shape, petiole wings broad, lamina large, upper surface dark green, glossy, slightly leathery, margins entire towards the base, apex rounded, base broadly to narrow rounded, aroma mild. Inflorescence cymose clusters, solitary flowers also in the axils of leaf, bisexual, 2-9 flowers in cluster, buds ovate oblong, white, minutely hairy at the apex, pedicel cylindrical, long, anther long. Fruit medium large, oblate to globose, colour light yellow, surface smooth, no ridges and furrows, apex evenly rounded, areole absent. Rind medium thick, pleasant strong aroma, segment 13, seeds average 39 per fruit, TSS 7.5%, acidity 0.85 and juice abundant.

Marsh Seedless

Best flavor if grown in hot climate locations. Fruits take 18 months from bloom to ripen. Tree has dense form and rich, green foliage. Fruits are big, smooth with white flesh. This is seedless cultivar.

Rio Red

Texas hybrid that shows much better color than the traditional 'Ruby'. Best in warmer locations but also performs well near the coast.

Star Ruby

Dark red color is characteristic of this variety, even when grown in cooler coastal regions. Plants are small with dense foliage. Heavy bearer, fruit are smooth with dark red flesh, stores well on the tree.

VI. Pummelo (Citrus grandis)

Pummelo, sometimes referred to as Shaddock, is a typically large sized tropical citrus fruit. Pummelo is probably native to southern China or the Malay and Indian archipelagos. A monoembryonic species with large sized fruits. Spreading, round topped, almost thornless tree. Leaves large with broadly winged petiole. Lower surface of leaf is pubescent, particularly main vein. Flowers very large, fruits large sized, subglobose to pyriform in shape with thick and spongy rind. Fruits are sweet and moderately juicy. Rind thick and seeds are very large.

VII. Sweet Lemon (C. *limetta)*

A general name for certain non-acid lemons or limettas. In India, they are grown in the Nilgiris, Malabar and other areas. The fruits are usually insipid, occasionally subacid or acid. The seeds are white within and the tree is large, resembling that of the orange.

VIII. Galgal or Hill Lemon (C. *pseudolimon*)

It is cultivated on a commercial scale in the submountain areas of northern India where it is grown as a substitute for lemon or lime. It is also known as Kumaon lemon. The notable differences between this variety and other lemon varieties are that it has the essential oils, aroma in both leaves and rind, single bloom and one crop behavior. It is commonly used for making pickle in Punjab. Tree fairly tall, hardy and vigorous, upright and spreading with an irregular and loose crown, foliage dense and light green thorny, leaves broadly elliptic ovate to oblong, marginate crenate, apex obtuse-acute, petiole long marginally winged. Fruit ovate oblong, yellow, apex slightly nipped, base rounded toi slightly nippled, rind medium thick, axis hallow, segments 8-10, pulp light yellow, juicy, flavour good and sour. Seeds are 28-60.

IX. Rootstock Varieties

Karna khatta, Sohsarkar (*C. karna*)

It is moderately plyembryonic species. Tree medium to large, some what similar to that of rough lemon. Leaves large with serrulate margin and winged petiole. Flowers are large and pigmented. Fruits medium, rind surface irregular and apical papilla well developed. Rind thick and moderately adhering. Fruit surface and pulp orange coloured. Fruit juicy and acidic in taste

Cleopatra mandarin (*C. reshni*)

Tree thornless, with dense top, leaves small, fruits produced singly or in bunches, fruit colour dark orange red shape ovalate, flattened at both sise, size small, rind rough, thin, loosely attached, flesh orange, juice abundant, acidity and sweetness normal.

Rough lemon (*C. jambhiri*)

Perhaps a lemon x citron hybrid, but has been given the botanical name of C. *jambhiri* originated in northern India, where it grows wild; fruits oblate, rounded or oval, base flat to distinctly necked, apex rounded with a more or less sunken nipple; of medium size, peel lemon-yellow to orange-yellow, rough and irregular, with large oil glands, often ribbed; pulp lemon-yellow, usually in 10 segments, medium-juicy, medium-acid, with moderate lemon odor and flavor; seeds small, 10 to 15, brownish. Reproduces true from seeds, which are 96% to 100% nucellar. Tree large, very thorny; new growth slightly tinged with red; buds and flowers with red-purple. The tree has been of great importance as a rootstock for the sweet orange, mandarin orange and grapefruit. Jatti khatti, Florida rough and Italian rough are the variants of rough lemon.

Attani *(C. rugulosa)*

The leaves are mandarin like but have larger wing. The rind peels easily. The flowers are medium sized with a pubescent ovary.

Rangpur lime *(C. limonia)*

Medium large, fairly vigorous growing tree, crown compact, trunk medium, robust, branches spreading, bark rather thin, brown, young shoots purple coloured, spines are on one year or two year shoots, no spines on trunk or older growth. Petiole medium long, cylindrical, articulated, slightly winged, lamina medium large, ovate elliptic, upper surface green, apex attenuated, base euneate, aroma distinct lemon like. Inflorescence cymose cluster, sub terminal or superiously terminal, proportion of bisexual flowers are more, flower bud small, light purple, ovate star shaped at the apex, male flowers generally 4 petalled, light purple out side and white in side, ovary, ovoid or sub- cylindrical. Fruit shape globose, oblate, colour deep orange, surface smooth, finely papillate, rounded or depressed, furrows extending through the color, stem short, apex rounded, areole distinct. Rind medium thick, oil medium abundant, aroma strong, measocarp medium thick, light orange, segments easily separable, juice abundant, TSS 5.5%, taste sour, acidity 3.18%. In reality this orange-colored fruit is a sour mandarin. Its juice combines mellow lime sourness with mild orange flavor.

Sour orange (*Citrus aurantium*)

A medium-sized tree, with a rounded top; twigs angled when young, with single, slender spines, often short, or stout spines on rapidly growing shoots; leaves medium-sized, ovate, bluntly pointed at tip, broadly rounded to cuneate at base; petioles 2-3 cm long, rather broadly winged, flowers large, very fragrant fruits subglobose, usually slightly depressed at both base and top, peel thick, with a rather rough surface, becoming brilliant orange with a reddish tint at maturity;

locules 10-12, filled with sharply acid pulp and numerous seeds; fruit becoming hollow at center as it matures.

Alemow *(C. macrophylla)*

It has large leaves with much smaller, subtriangular, short-winged petioles. The fruits are very large, 8.5 to 10 cm in diameter, subglobose to oblong, more or less narrowed at the base, with a rough, transversely-corrugated, but rather thin skin. The fruit has 13 to 16 segments and rather dry, sour pulp, considered inedible.

Citron *(C. medica)*

A shrub or small tree, with a short indistinct trunk, rather spiny, young shoot smooth, leaves large, not articulated oblong ovate, elliptic in shape, margins serrate, petiole wingless, flower in cymose clusters, percentage of stamens flowers are more, fruit lemon yellow, large 10-20 cm long oblong elluiptyical, rough or warty, rind very thick, pulp sparse, juice moderate, acidity strong, seeds numerous.

Gazanimma *(C. gaznima)*

A distinctive species having crushed leaf aroma similar to ginger or eucalyptus smell. Medium sized tree with thick glossy leaves. Fruit medium sized, almost spherical and smooth surfaced. Fruits are in acidic taste

Adajamir (C. assamensis)

A distinctive species having crushed leaf aroma similar to ginger or eucalyptus smell. Medium sized tree with thick glossy leaves. Fruit medium sized, almost spherical and smooth surfaced. Fruits are in acidic taste

Calamondin *(C. madurensis)*

Tree small topped ornamental with upright branches, columnar, rather bushy and dense, slightly thorny, leaves broadly oval, lighter green below, petiole short, narrowly winged. Flowers white small, borne singly or in pairs at the ends of branchlets. Fruit colour orange to deep orange, surface smooth and glossy, very finely pitted, shape oblate to spherical, size small, base flattened, rind thin, loose, easily separable when ripe, tender and pulp very acidic of good flavour, seeds less. Prized fruit of the Philippines, know as 'Kalamansi'. Small orange fruit, sour in taste, can be used as a lime or to make marmalade.

Indian wild orange *(Citrus indica)*

Branches terete, spiny, glabrous; leaves oblong or lanceolate, thick, acute at the base, veins curved, petioles articulated, linear, fruits small, broadly obovoid or sub-pyriform, solitary on terminal twigs, pedicels very short; segments few, vesicles fusiform; seeds large, smooth, mono- embryonic. This species has leaves resembling those of *C. sinensis*, but small, fig-shaped fruit containing extremely large seeds is entirely different from any other *Citrus*.

Papeda

Khasi papeda (*C. latipes*)

A thorny tree similar to inchang papeda but having leaf blade more variable in size and shape and with tips subacute. Flowers borne in small axillary racemes. Inflorescence is recemose, flower bud medium sized Fruit are borne singly having thick peel; inner layer is chalky white just below the outer green layer. Seeds are smaller.

Microptera (C. microptera)

Petiole broadly winged, distinctly articulated, leaves elongate acuminate, twice as long as petiole, twigs sub compresses with long one spine on young twigs and one short spine on old twigs in the axil of the leaves. Fruit with 10 – 12 segments, pubescent with scanty pulp, depressed, almost without juice. Fruit globose and pale yellow.

Ichang papeda (*Citrus ichangensis*)

A spiny shrub or small tree, twigs angular when young, with stout, sharp spines, leaves narrow, petioles very large, broadly winged, obovate or oblong-spatulate, evenly rounded at the tip and narrowed abruptly at the base, leaf blade ovate-acuminate, flowers 2.5-3 cm diameter with minutely ciliate margins; petals oblong, stamens 20, style very short, caducous; stigma nearly as large as the ovary; ovary with 7-9 locules, ovules numerous in each locule; fruits small, glabrous, peel rough, seeds large, very thick, apparently monoembryonic.

Trifoliate orange (*P. trifoliata*)

It is small much branched tree, twigs angled, spiny, spines single, stout, straight, sharp, buds covered with overlapping bud scales. Leaves palmately, trifoliate, flowered bud covered with scales, flower single, sessile, fruits small, almost sessile, globose or ovoid or slightly pyriform, dull lemon colour, peel thick, oil glands numerous. *Poncirus* is widely used as a rootstock; A rather dwarf form named 'Flying Dragon', has been recently experimented with as a rootstock.

Fortunella

Small orange fruit that are eaten peel and all. Will store on trees for months without loss of flavor. Needs lots of heat to produce very fragrant blossoms in summer. Plant is very cold hardy. Native of China, it is a symbol of prosperity and good luck. *Fortunella* differs from Citrus mainly in having two collateral ovules near the top of each locule (*Citrus* has 4-12). Though evergreen, it possesses some degree of winter dormancy, enabling the tree to remain quiescent during the weeks of warm weather. *F. margarita* and *F. japonica* are quite widely cultivated in China, Japan and some subtropical environments. Fruit has a relatively thick, fleshy sweet and edible peel, and 4-7 segments filled with pleasant, mildly acid pulp.

Clymenia

This genus is allied to *Citrus*; it differs in with citrus in flowing aspects

1. Subsessile pulp-vesicles (somewhat narrowed at the base but not borne on slender stalks) attached in great numbers to the lateral segment walls for 3/4 the distance from the peel to the axis;

2. Very numerous stamens (50-100) with free, slender filaments

3. Ovary with a very short, stout style

4. Leaves with very short petioles,

Small trees, branches spineless, twigs subangular when young, then cylindrical; leaves thin, smooth, acuminate-caudate at apex, cuneate at base, tapering into the very short petiole, furrowed above, and not articulated with the leaf blade. The flowers arise singly in the leaf axils and are borne on straight, rather stout pedicels, slightly longer than the petioles of the subtending leaves; calyx persistent, stamens very numerous, filaments free, slender, ovary ovoid with 14-16 locules, ovules several in each locule; style very short; fruit ovoid, small, skin thin, orange-like.

Citrus Hybrids

Citranges

These are group of hybrids. The hybrid showed intermediate characters of the parents. The leaves are manly trifoliate evergreen. The fruits are yellow to orange in colour and rind is thin smooth. The fruits are juicy flavoured.

Troyer citrange Sweet orange x Trifoliate orange

Morton citrange Sweet orange x Trifoliate orange

Malta

Mosambi

Kagzi lime

Kagzi Kalan

Galgal

Sweet lemon

Red Blush

Marsh Seedless

Calamondin

Attani

Carrizo citrange

Attani

Jatti Khatti   Rangpur lime

Trifoliate orange

3. Grape

Varieties Identified from IARI, New Delhi

Pusa Seedless

It is a selection made from Thompson Seedless. It resembles Thompson Seedless with regard to most of the characters but has elongated berries. It is highly responsive to GA3 application. The berries have high T.S.S. and suitable for both table purpose and raisin making.

Beauty Seedless

It was introduced by this Division from California, USA ripening 4th week May. The vine medium, bunches medium to large, long shouldered and compact with bluish black coloured, spherical; berries medium, prolific bearer, keeping quality very low.

Pusa Navrang (Madeleine Angevine X Rubired)

This hybrid has been released in 1996 from the division of Fruits and Horticultural Technology, IARI. It is an early ripening (1st week of June), basal bearer, teinturier variety containing red pigment both in peel and pulp. The bunch is loose, medium with round and medium berries; suited for coloured juice and wine making; resistant to anthracnose disease.

Pusa Urvashi (Hur x Beauty Seedless)

This variety has been released from IARI, New Delhi, basal bearer, ripening during 1st week of June. Bunch is loose and medium in size with seedless greenish-yellow berries. It is suitable for table purpose and raisin making. The pulp T.S.S. varies from 20 to 22%.

Varieties identified from IIHR, Bangalore

Arka Shyam (Bangalore Blue x Black Champa)

Moderate to heavy yielder, Medium clusters, berries big, round; TSS 24 °Brix, acidity 0.6%, suitable for table and wine purpose.

Arka Hans (Anab-e-Shahi x Bangalore Blue)

Prolific bearer, bunches medium, yellowish green berries, seeded, TSS 21°Brix, acidity 0.5%, suitable for wine making.

Arka Sweta (Anab-e-Shahi x Thompson Seedless)

Prolific bearer, suitable for table use and raisin making, bunch weight 260 g, berries greenish yellow, obovid, uniform, seedless; berry weight. 4.08 g, TSS 18-19° Brix, acidity 0.5-0.6%.

Arka Majestic (Angur Kalan x Black Champa)

plants vigorous, prolific bearer, high yielder, berries deep red, obovoid, bold and seeded, suitable for table use, bunch weight 370 g, berry weight 7.7 g, TSS 18-20° Brix, acidity 0.40.6%, pedicel attachment very good, ideal for export, all buds are fruitful.

Arkaneelmani (Black Champa x Thompson Seedless)

plants vigorous, well filled to slightly compact bunches weighing on an average 360 g, berries black, seedless, berry wt. 3.2 g; TSS 20-22° Brix, acidity 0.6-0.7%, all buds on a cane are fruitful.

Arka Chitra (Angur Kalan x Anab-e-Shahi)

Prolific, high yielding, berries-golden yellow with pink blush, seeded but attractive, suitable for table purpose, average bunch weight 260 g, berry weight 4 g, TSS 18-19° Brix.

Arka Soma (Anab-e-Shahi x Queen of vineyards)

Heavy yielder, white berries, seeded, meaty pulp with muscat flavour, good for wine making.

Arka Trishna (Bangalore Blue x Convent Large Black)

Prolific bearer, berries deep tan, seeded, very sweet pulp, male sterile, good for wine making

Arka Krishna (Black Champa x Thompson Seedless]

Prolific bearer, berries-black, seedless and sweet, more juicy and suitable for beverage industry

Other varieties

Anab-e Shahi

The origin of this cultivar seems to be a bud sport as it had a satellite chromosome. It has been acclaimed as one of the most productive cultivar grown in India and yielded over 30-40 tones/acre/year. It has attractive large bunches of berries with good shipping quality.

Bangalore Blue

It is a hybrid of *vinifera* X *labrusca.* Vine medium and moderate yielder, does well on Kniffin as well as bower systems of training. The bunches small, compact; berries small to medium, dark blackish purple. The ripening uniform. Apart from being used for table purpose, it is being extensively used for juice and wine making. It is known for its hardiness and resistance to disease for which it finds a suitable place as parent in a breeding programme aimed at inducing disease resistance.

Perlette

This cultivar is a hybrid of Scolokertekhiralynoje 26 X Sultanina Marble developed at the University' of California, Davis, by Dr. H. P. Olmo. The most striking feature is the transluscence of the mature fruit. Berries medium, whitish green, spherical; flesh soft, mild, muscat flavoured. It has good keeping quality. Small underdeveloped berries (shot berries) scattered all over the bunch is major defect. The bunch compactness is reduced in a new strain „Loose Perlette' which was the result of irradiation.

4. Papaya

Varieties released from IARI, New Delhi

Pusa Delicious:

It is a gynodioecious line. The plant medium; first bearing starts at the height of 2m; fruits medium (1.15 to 2.5 kg), flavour; pulp bright orange, T.S.S. ranges from 10 to 13° Brix.

Pusa Majesty

Gynodioecious line, good yielder. The fruits have good texture and firm flesh enabling long distance transportation. Fruit quality is very good with very high papain content. The variety is also resistant to nematodes.

Pusa Dwarf

Plant dwarf, fruits oval round and medium (1 to 1.5 kg). It starts fruiting at a height of 35cm above the ground level. This varies ideally suited for high density planting.

Pusa Nanha

This variety was evolved through mutation. It is exceptionally dwarf (1 m); fruiting starts at 30cm height from ground level. It is suitable for planting in pots and under high density planting systems.

Pusa Giant

This is a vigorous variety with large sized fruits. The plants are sturdy and resistant to strong winds. It starts bearing at a height of 90 cm from ground level.

Varieties Released from Tamilnadu University, Coimbtor

Co. 1

It is a selection from the progenies of cv. Ranchi. Plant dwarf, fruit round or oval, golden yellow skin, flesh orange.

CO-2

This is a pure line selection from a local type. The plant medium-tall, fruits obovate, large; skin yellowish green; flesh orange, soft, moderately juicy. It is a good table fruit and also a high papain yielding cultivar.

CO-3

This is a hybrid between Co. 2 X Sunrise Solo, a tall vigorous plant. The fruit medium, sweet keeping quality good.

CO-4

This was evolved from a cross between Co. I X Washington. Plant medium tall, fruit large, flesh thick, yellow with purple tinge, taste sweet; good keeping quality,

CO-5

It is a selection from Washington and found good for papain production.

CO-6

It is a selection from Pusa Majesty. Plant is dioecious and dwarf and found suitable for papain production.

Other Varieties

Coorg Honey Dew

It is a chance seedling of Honey dew. Plant dwarf, heavy bearer, mostly hermaphrodite but occasionally pistillate flowers. Fruits are oblong and flesh is thick with good flavour.

Punjab Sweet

It is a selection made at PAU, Ludhiana. It is frost tolerant and dieoecious in nature.

Surya

It is a gynodieocious hybrid (Sun rise solo x Pink Flesh sweet) developed at IIHR, Bangalore. It is high yielding variety with good quality fruits.

HPSC-3

It is a hybrid between Tripura Local x Honey Dew and found resistant to papaya mosaic virus.

RCTP-I

High yielding gynodiocious lines developed through selection, from ICAR, Tripura Centre, Lembucherra. Plant tall with single straight stem without branching. Fruits well shaped adequately spaced. A plant produced an average fruit yield of 63.25 kg. The total soluble solids 12.5° brix.

Pusa Nanha   Pusa Dwarf   Red Lady

T-1   CO-5

5. Banana

Poovan

The plant tall, hardy, grows vigorously under the ratooning system of cultivation. One of the distinguishing characters of the plant is the rose-pink colour on the outer side of midrib. It can grow under unirrigated condition or with scanty irrigation. The fruit medium to small, yellow skinned, flesh firm with a sub-acid taste, good keeping quality, resistant to Panama wilt and fairly resistant to bunchy top. The average bunch weight is about 15 kg.

Dwarf Cavendish

It is the leading commercial cultivar of Maharashtra. The plant dwarf, fruits large, curved, peel thick, greenish, flesh soft, sweet. The greenish colour of the fruit is retained to some extent even after ripening, Susceptible to bunchy top and leaf spot disease but resistant to Panama wilt, keeping quality not good. The Dwarf Cavendish is an important banana in international trade. Two semi-tall mutants, Monsmari and Williams Hybrid are widely grown in Queensland, Lacatan.

Kanchkela

This is the most important commercial culinary banana cultivar of India. The plant tall, robust, light green, very hardy and grows under un-irrigated condition. Average bunch weight is about 15 kg.

Harichal

Musa (AAA) group-Syn. Bombay green (Maharashtra), Peddapachaarati (Andhra Pradesh), Robusta (Tamil Nadu). It is a semi-tall sport of Dwarf Cavendish, It is another important commercial banana of Maharashtra. Fruits large, peel thick, greenish to dull yellow, sweet and delicious. The fruits have better keeping quality than that of Dwarf Cavendish. Average bunch weight is about 20 kg.

Martman

It is the choicest table cultivar of West Bengal. The plant is tall and can be identified by the yellowish green stem with brownish blotches, reddish margins of the petiole and leaf sheath. The average bunch weight is about 12 kg. Fruits medium, similar to that of Poovan in appearance, peel thin, ivory yellow, flesh firm, sweet with a pleasant aroma. Its cultivation is decreasing due to susceptibility to Panama wilt.

Nendrun

Musa (AAB) group-Syn. Ethakai (Kerala), Rajeli (Maharashtra), Kochikehel *(Sri* Lanka), Plantain (Trinidad). This cultivar is known in all part of the world as plantain. This is a dual- purpose cultivar of Kerala. It has very good keeping quality. The fruit is relatively longer and thicker than most other bananas. The bunch is not compact. The average bunch weight is 15 kg.

Safed Velchi

This variety is under stray cultivation throughout South India and Maharashtra and mostly grown as intercrop in coconut and arecanut garden. The plants are medium-sized with slender, yellowish pseudo stem having reddish petiole margin. The fruits are small, firm-fleshed and sweet. The average bunch weight is about 12 kg.

Gros Michel

Gros Michel occupied the first rank in desirable fruit characters, such as size, quality, flavour, attractive skin colour, resistance to bruising, grade yield, symmetry and strength of bunch. The most serious demerit of this cultivar is its susceptibility to wilt.

6. Guava

Allahabad safeda

Fruit medium, roundish, average weight 171 g; surface smooth, glossy, peel yellowish white with small distinct dots; flesh white, soft melting; flavor pleasant; sweet, good for canning.

Apple colour

Fruit medium, spherical, roundish, fruit weight 38 g; surface slightly rough; peel thick, light red or pinkish with scattered dots; flesh creamy white, soft melting; flavor pleasant, sweet.

Chittidar

Fruit small to medium, roundish ovate, average fruit weight 126 g; surface smooth, glossy, peel straw-yellow with few scattered dots; flesh yellowish white, soft melting, flavour mild, sweet acidic, good for canning.

Behat coconut

Fruit large, elliptical round, average weight 270 g; surface slightly rough, peel thin, light greenish yellow; flesh white, crisp; flavour mild, acidic sweet; seed large.

Lucknow 49 (Sardar)

Fruit medium to large terminate to pyriform; average weight 182 g; surface slightly rough; peel thick, yellowish white, distinct medium dots; flesh creamy white, soft melting,; flavour slightly acidic, swett, good for canning and jelly.

Lalit

It has been released by the CISH, Lucknow, for commercial cultivation. The fruits medium (185 g), attractive, saffron-yellow, red blush; flesh firm, pink with good blend of sugar and acid. This is suitable for both table and processing purposes.

7. Ber

Thar Sevika

Developed from a cross between Seb x Katha. It is an early maturing variety. Plants semi- spreadin; fruits attractive, greenish yellow, juicy; average fruit weight 25 g; stone weight 1.35 g, flesh thickness 1.1 cm. TSS 24%, total sugar 5.02%, ascorbic acid content 88 mg/100g fruit and acidity 0.60. it is free from powdery mildew and have low incidence of *Alternaria* rot. Ripening during last week of December.

Thar Bhubhraj

It is a selection from local material of Bhusavar area of Bharatpur district of Rajasthan, It has an upright growth habit and has the ability to withstand extremes of temperatures of 2.5 to 48.0° C. It is also an early maturing variety. Fruits are attractive and very juicy, fruit weight 27.0 g, stone weight 1.9 g, flesh thickness 1.0 cm, TSS 23%, total sugar 4.29%, ascorbic acid content 60 mg/100 g fruit, acidity 0.80%. The fruits mature by end of December-first week of January.

Umran

The fruits large, oval in shape, with roundish apex. The fruit matures during February-March and ripens during mid March. Fruits are sweet, TSS 19%, acidity 0.12%. This cultivar has long shelf life.

Gola

It is an early mauturing cultivar. Fruits are very attractive, roundish in shape and golden yellow colour. Fruit weight 20.0 g, TSS 17-19%, acidity 0.46-0.51%.

Thar Sevika   Thar Bhubhraj

8. Apple

Red Delicious

The trees vigorous, form spur freely. Fruit large and oblong-conical, with red streaks. Flesh creamish, juicy, aromatic, sweet. It ripen-in the third week of August and can be stored for 3-4 months.

Starking Delicious

The trees vigorous, form spur freely, fruit large, oblong conical in shape; ground greenish-yellow covered with dark red stripes all over the fruits. Flesh creamish, juicy, aromatic, sweet, quality excellent. The fruits cannot be stored for a long time. They ripen in second week of August.

Richared

The trees vigorous, form spur freely, fruits large, oblong-conical; ground colour greenish-yellow, covering with red wash all over. Lenticells are conspicuous. It also ripens in third week of August.

Ambri

This is the only indigenous variety grown in India. It originated in Kashmir perhaps from a seedling. The trees vigorous, .fruit medium to large, oblong ; red streaks over a greenish-yellow background. The pulp white, crisp and sweet. The fruits ripen in the last week of September. They can be stored for 4-5 months under ordinary storage in Kashmir and for 10 months in cold storage. It is an attractive apple with an extra-ordinary keeping quality.

Baldwin

Discovered as a chance seedling, the trees moderately vigorous, spreading, producing spurs freely. The fruit large and round to oblong, ground colour pale-green, flushed with dull purplish brown. Sometime traces of red stripes are present. The fruits mature in the beginning of August. The fruits last 34 months under ordinary storage conditions. The tree bears late in life and is a biennial bearer. It is a triploid variety. It was grown extensively in Kullu valley, but with the introduction of Red Delicious group, it has almost been completely removed.

Golden Delicious

This is most popular cultivar of all apple-growing areas of the world. In India, it is commercially used as a pollinizer for Delicious apples. It is not popular in Indian market as a commercial cultivar because of its yellow colour. However, some orchardists are getting good price from it. Its trees are moderately vigorous, spreading, producing spurs freely. Fruit medium and round conical to oblong.

Ground colour greenish yellow which turns to golden-yellow on ripening. Some fruits are halfflushed with pale orange. Flesh is creamish, crisp, fine textured, juicy, sweet, little acidic with very good aromatic flavour.

Granny Smith

It is a late season variety. A chance seedling from the backyard of Marie Ann Smith, Australia. The tree is very vigorous and crops heavily, but it is not much good for areas with short growing seasons. Fruits green, slippery skinned, dual purpose cooking/eating. The flesh is hard, crisp, and juicy. The flavor is tart, becoming very sweet if tree ripened. The fruit will store for several months after maturity without needing refrigeration. It. is an excellent pollen source for other varieties.

Jonathan

It is mid season variety. It is a well-known cultivar throughout the apple-growing areas. The trees moderately vigorous, producing spurs freely, precocious and regular-bearer. The fruits medium and round to round-conical, ground colour pale greenish-white and flushed with bright crimson and scattered broken red stripes. The flesh is white, slightly acidic blend. The fruits can be stored for 2-3 months. The fruit is often affected by black freckles, known as 'Jonathan Spot'. It also acts as a pollinizer for Delicious varieties.

Vance delicious

A mutant of Delicious. Trees vigorous, spreading; fruit medium to large, conical, striped red; flesh greenish, turning to creamish-yellow on ripening; firm, juicy, aromatic and sweet. The fruits ripen 12-13 days earlier than Starking Delicious. It develops good colour at low altitudes and marginal areas. At higher altitudes it may over colour and turn to dark blackish-red. It is more productive than Starking Delicious.

Jonagold

A late season variety. It is developed by crossing of Jonathan and Golden Delicious. The fruit are striped red over a yellow ground color, fine textured, juicy, and are sweet and with a bit more acidity than Golden Delicious. Consistently rated as one of the finest culinary apples. The fruit are usually large. Requires a pollenizer (self infertile due to being triploid). Very vigorous and with a spreading growth habit,

Spur Type Varieties

Red Chief

This is a limb sport of Starkrimson. Tree size small, compact and forms more numberof spurs. Fruit medium to large, uniform, conical, dark and stripes are present on blush.. It develops colour 15 days earlier than Starking Delicious. Flesh is creamish yellow, firm, juicy, aromatic and sweet in taste

Red Spur

The trees are two-thirds to those of standard Delicious cultivars. Trees have close internodal growth. Fruit medium to large, conical , dark red. They resemble to those of Richared. Flesh creamish-yellow, firm, juicy, sweet . It matures 2 weeks earlier than Starking Delicious

Oregon Spur

Tree size is two-thirds the size of standard Delicious cultivar. It forms spurs heavily. Fruit is medium to large, conical in shape and blushed with dark red colour. Fruit may be uniformly coloured. Flesh creamish-yellow, firm, juicy and sweet in taste. The fruits ripen a few days earlier than Starking Delicious in Kullu valley. It is recommended for higher altitudes.

9. Pear

Bartlett

It is also an important mid season, ripening start from the first week of August. The fruit big, obtuse, pyriform, yellow; peel thin, pulp soft, juicy, sweet and scented. The keeping quality of the fruit is poor. It is a standard usually requires thinning. The fruit is fit for canning.

Doyenne Du Cornice

It is another mid season variety which requires shorter period of chilling. The pulp is melting, delicious and aromatic. The fruit is big and obtuse pyriform, the colour of which is yellow with russet patches. The stalk is long and thin. It had been reported that the variety gave good response when planted with 'Winter Nelis' as pollinizer. It is irregular in nature. The keeping quality of the fruit is poor.

Beurre Hardy

It is a late variety. The fruit roundish oval and big; grayish golden yellow; pulp soft, juicy, sweet and aromatic. Fruits are picked by second week of August. It is also grown in South India above 1,700 metres.

Star King Delicious

Fruits big, pyriform, golden yellow; flesh creamy yellow, soft, juicy, sweet, aromatic, and delicious. Fruits are picked at the end of August.

Patharnakh

Late season variety of Chinese group. The fruit light yellowish grey with prominent brown dots; flesh white. Average fruit weight at maturity ranges between 200-250 g. Juice content is 45-50%, TSS 12 to 15%, acidity 0.30 to 0.35%. Keeping quality of the fruit is very good. Picking starts from the 2nd week of August. The fruit is very hard and quite fit for long distance transportation.

Baggugosha

It is a most important variety of Kashmir valley and can also be grown successfully in submountaneous tracts, but the quality becomes poor. Fruit weight 110 to 205 g, juice content 43 to 59%, TSS 12.9 to 17.2%, acidity 0.23 to 0.44%. It is an interspecific hybrid and requires to be improved by adding some genes responsible for regular bearing.

Kieffer

It is cultivated extensively in the Kodaikanal area in South. It is a cross between the French pear and the Oriental pear and originated in America. Fruit are large, showing different shape on the same tree, yellow when fully ripe. Flesh is soft, slightly gritty, very juicy and sweet. A prolific bearer variety which ripens in September.

Flemish Beauty

The tree vigorous, very productive; fruit high in quality, but require careful timing of harvest to obtain full flavour and freedom from breakdown. Flemish Beauty is susceptible to scale and fire blight.

2

Propagation of Fruit Trees

The propagation of plants has been one of the fundamental occupations of mankind since civilization begins. Most of the selected plant would have lost or could have attained undesirable forms if they were not propagated under strict environment conditions. Growth, in all nature, result from the division of cells to form more similar cells. Thus bacteria reproduce by simple division, in higher plants and animals, a new individual is genetically formed only when cell division fallows the fusion of two cells. This is known as sexual reproduction. There are thus two methods of producing new plants, sexual and asexual or vegetative reproduction.

I. Sexual Propagation

Raising of plants by means of seeds is called sexual propagation. There are many plants raised for our garden from seeds. The plants produced through seeds are called seedlings. In sexual method, the sex organs of flower are involved in process like pollination and fertilization, resulting the formation of seeds. eg rootstocks plants of many fruit crops are grown through seed propagation.

II. Asexual Propagation

1. Propagation by apomictic seedling

   eg mango, apple and citrus

2. Propagation by vegetative structure

   1. Propagation of division

      Division of rhizome

      Banana, blueberry etc.

   2. Propagation by suckers

      Pineapple, banana, blackberry, raspberry etc.

   3. Propagation by runners

      Strawberry

4. Propagation by offset

Pine apple, banana etc.

Propagation by Cutting

A. Stem cutting

Stem cuttings are the most convenient and popular method of plant propagation. A stem

cutting is any cutting taken from the main shoot of a plant or any side shoot growing from the same plant or stem. It is essential for the cuttings to have a sufficient reserve food to keep tissue alive until root and shoot are produced. The shoots with high carbohydrates content roots better. Cuttings from young plants root better, but, if older shoot of the plant are cut back hard, very often they can be induced to produce suitable shoots for rooting. There are several types of stem cuttings.

1. Hardwood cutting

Hard wood cuttings are prepared during dormant season, usually from one year old immature shoots of previous season s growth. The length of cuttings varies from 10-75 cm in length and 0.5 to 2.5 cm in diameter, depending upon species. Each cutting should have at least two buds. While preparing the cutting, a straight cut is given at the base of shoot below the node while a slanting cut 1-2 cm above the bud is given at the top. eg Grape, fig, pomegranate, mulberry, kiwifruit, olive, quince, hazel nut, chest nut, plum, gooseberry and apple.

2. Semi hardwood cutting

Semi hardwood cuttings are prepared from partial matured, slightly woody shoot. These are succulent and tender in nature and are usually prepared from growing wood of current season s growth. The length of cutting varies from 7-20 cm. The cuttings are prepared by trimming the cutting with straight cut below a node and removing a few lower leaves. However, it is better to retain two to four leaves on the top of cuttings. Treating the cutting with 5000 ppm IBA before planting gives better results. eg Mango, guava, jackfruit, lemon.

Softwood cutting

Softwood cutting is the name given to any cuttings prepared from soft, succulent and non-lignified shoots which have not become hard or woody. Usually the cutting size is 5-7.5 cm but it varies from species to species. Usually some leaves are retained and before planting, treatment with auxin (IBA) is beneficial. eg lime and lemon.

B. Leaf bud cutting

Leaf cutting should preferably be prepared during growing season bacuase buds if inter in dormancy may be difficult to force to active stage. A leaf bud cutting consists of a leaf blade, petiole and shoot piece of stem with attached axillary bud of active growing leaves. In this cutting, 1-1.5 cm stem portion is used when propagating material is small. eg. Blackberry, lemon, raspberry.

C. Root Cutting

Blackberry, fig, cherry, raspberry

Propagation by layering

1. Tip layering

In tip layering, the tip of shoots is bend to the ground and the rooting takes place near the tip of current season s shoot. The stem of these plants completes its life in two years. The tips of shoot buried 5-10 cm deep in the soil. Rooting in the buried shoots takes place within a month. The new plants may be detached and transplanted in the nursery during spring. eg Black berry, raspberry.

2. Simple layering

In actual practice of tip layering, the flexible shoots of a plant are bent downwards over to ground in early spring or in rainy season. A second bend in the shoot, a short distance from the tip, which is covered with soil and held in place with wire or wood stakes. This portion is sometimes injured by notching and girdling to stimulate rooting. eg Grape, lemon.

3. Trench layering

In this method it is important to establish a permanent row of plants to be propagated. The mother plants are planted at the base of a trench at a angle of 45° in rows. The long and flexible stems of these plants are pegged down on the ground to form a continuous line of layered plants. The young shoots then arised from these plants are gradually mounded up to a depth of 15-20 cm in autumn, winter or end of the season, depending upon the species. eg Apple rootstocks (M16 and M25), cherry, plum

4. Serpentine layering

It is modification of simple layering in which one year old branch is alternatively covered and exposed along its length. The stem is girdled at different point in the underground. The stem is girdled at different points in the underground part. However, the exposed portion of the stem should have at least one bud to develop a new shoot. After rooting, the section are cut and planted in the field. eg American grape.

5. Air layering

Litchi, Lime and sweet lime can be propagated by air layering. Generally one to two years old shoots are used for air layering. First the leaves are removed from the base of the selected shoots then the stem is given a notch or is girdled by removing a ring of bark about 2-3 cm wide. Application of root promoting hormones at the time of layering helps to get profuse rooting within a short time. Root promoting substances may be applied as powder or in lanolin or as a solution. IBA at the rate of 500-1000ppm or a combination of IBA + NAA, both at the rate of 500ppm may be applied for better results. After application of hormones, ringed or girdled area is then covered with handful of moist soil. This ball of earth should be again covered with sphagnum moss and wrapped with a polythene sheet. Air layering should be done either in spring or in mansoon. The rooted layers are either planted in pots or in the nursery beds in a shady place untill they are fully established.

6. Stooling/mound layering

In this method the mother plants are headed back to 5-10 cm above ground level during dormant season. The new sprout will arise within two months. These sprouts are then girdled and rooting hormone made in lanolin paste is applied to the upper portion of the ring. The concentration of rooting hormones are varies from plant to plant but in general 3000 to 5000 ppm is most commonly used. These shoots are left for two days for proper absorption of hormone before they are covered with soil. Care must be taken to keep the soil most. The roots from shoots may emerged within 30-40 days. These rooted stools should be severed from the mother plant only after 60-70 days and then planted in the nursery beds. eg Apple rootstocks, guava, mango, litchi.

Propagation by Grafting

In grafting, the stock and scion are placed in close contact with each other and held together firmly, until they unite to form a composite plant.

1. Whip grafting

It is simple and popular method of grafting. In this method of grafting, it is essential that both stock and scion should be of equal diameter. About one year old rootstock is headed back at a height of 23-25 cm from the soil and a diagonal cut is made at the distal end of the rootstock. A similar slanting cut is made on the proximal end of the scion. The cut surface of both rootstock and scion are bound together and tied firmly. Many fruit plant are propagated by whip grafting.

2. Cleft grafting

It is particularly suitable in rootstock having diameter greater than the scion. Rootstock with 5-7 cm or more girth is selected for this purpose. The rootstock is cleft grafted after decapitating the stock 45 cm above the ground level. The beheaded rootstock is split to about 5cm deep through the center of stem. After that a hard wooden wedge is inserted to keep open for the subsequent insertion of scion. The scion of 15-20 cm size is taken from a terminal shoot, which is more than three month old and then it is wedge securely (6-7 cm). The cleft of the scion then slipped into the split of the stock. In thicker rootstock more than one scion should be inserted. The graft should be thoroughly waxed to prevent wilting. eg Avocado, apple, pear, plum, mango.

3. Bark grafting

It should be done in spring when bark of the stock slips easily. In is important that scion used in bark grafting should be dormant. The stock is first sawed off at a point, where bark is smooth. Bark is split downward, about 5 cm from the top. Scion of 10-12 cm long, containing 2-3 buds are collected from the dormant wood and are preparing by giving slating cut (5cm) downward along one side of the base. The scion then inserted in the center of split between the bark and wood of the stock. The scion is kept firmly by using adhesive tape. eg Many fruit plant.

4. Inarching

It is generally used for repairing or replacing damaged root system and hence also called as repair grafting. Selection of parent tree for taking the scion is an important factor for its success. The scion plant should be healthy, vigorous and high yielding. The stock is brought close to the scion. A thin slice of bark (6-8 cm long and about 1/3 inch in thickness at height) at about 20 cm above the ground level is removed from the stock with a sharp knife. A similar cut is made in the scion. Thus the cambium layers of both stock and scion are exposed. These cuts are brought together and tied firmly with the help of alkathene strip. After successful union, stock above and scion below the graft union are looped of gradually. It is done soon after rainy season provided that temperature of the localities does not fall below the 15 °C. eg Mango, sapota, guava, litchi.

5. Veneer grafting

It is simple method of propagation and can be used in one year old rootstock seedlings having a diameter of 1.0-1.5 cm. For veneer grafting, 3-6 months old scion shoots with lust green leaves are selected. Usually, the terminal and next to terminal shoots are most ideal. The shoots are defoliated 5-10 days prior the grafting leaving the petiole attached. The rootstock is prepared by making a slating cut (5cm long) and an oblique cut is made at the base of first cut so that a piece of wood along with bark is removed. The base of the scion wood is then fitted into the rootstock in such a manner that the cut surface including the cambium layers of scion and rootstock face each other. The rootstock and scion are tied together with polythene tape. When scion growth begins the shoot of rootstock is removed above the graft union. eg Mango

There are other methods of grafting are also used for propagation of fruit plants.

6. Epicotyl grafting

Mango

7. Top working

Mango, ber, cashew nut, mulberry

8. Soft wood grafting

Cashew nut

9. Bridge grafting

For repairing of damaged fruit plant

Propagation by Budding

1. Shield or T- budding

One–year-old rootstock seedling of 25-30 cm height and 2-2.5 cm thickness is selected.

The bark of seedling should slip easily. For actual operation, a „T shaped cut is made on the selected portion of the stock with the help of sharp budding knife. The two flaps of bark are then loosened slightly with the help of budding knife. From the bud wood, which is selected from a healthy shoot of a current season s growth, the buds of middle portion are selected. These are removed from the bud wood by cutting shallowly about 5-6 mm below and 2-3 cm above the bud. This shield piece containing a bud is inserted carefully in „T shaped incision made on the stock. This bud then presses firmly and tied with polythene strip. After the bud has sprouted, the stock is cut to about 10-15 cm above the bud. eg citrus, plum, peach, cherry, ber, rose etc.

2. Patch budding

A rectangular patch of bark is removed completely from the stock and replace with a similar patch of bark containing a bud of desired variety. It is successfully used in species having thick bark such as walnut, pecan nut.

3. Chip budding

Chip budding is successful method of budding when the bark of the stock does not slip easily. A chip of bark and wood is removed from the smooth surface between the nodes of the stock. A chip of similar shape and size is then removed from the bud wood of desired cultivar. For which, a 2-3 cm long down ward cut is made through the bark and slightly in to the wood of the stock. Then a second cut of about 2.5 cm is made so that it bisects the first cut at an angle of 30-45°. in this way the chip of wood is removed from the stock. The bud chip then slipped in the place of rootstock from where chip has been removed. eg Mango, grape.

4. Ring budding

In ring budding, a complete ring of bark is removed from the stock and it is completely girdled. A similar ring of bark containing a bud is removed from the bud stick and is inserted on to the rootstock. In this budding both scion and stock should be of same size. It is utilized in peach, plum, ber, mulberry etc.

Seedlings of mango (Kurukkan)   Mango Rootstock in the bed

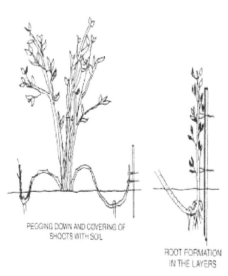

PEGGING DOWN AND COVERING OF
SHOOTS WITH SOIL

ROOT FORMATION
IN THE LAYERS

Simple layering

T- budding

Guava stool bed

Rooted stool of guava

3

Rootstock Quality and Important Rootstocks of Fruit Crops

Rootstocks play an important role in the propagation of plants. It may modify form or stature, adapt a variety of a soil, fit in an incompatible climate, impart or resist disease to the scion, increase production, hasten maturity of crop, change the colour of the fruit, effect the flavour of the fruit, shorten life of the tree, increase the size of the fruit, develop vigour, effect salt tolerance and influence storage capacity.

Quality of Good Rootstock

- The rootstocks should possess the following qualities:
- The rootstock must exhibit a high degree of congeniality with the scion variety and give maximum economic life to the tree.
- It should be well adaptable to the agro-climatic conditions of the proposed area.
- It should be resistant to diseases (mostly soil borne) and pests prevailing in the proposed area.
- It must exercise favourable influences on the performance, bearing and quality of fruits of scion.
- The stock should also have other desirable qualities like salt tolerance, drought resistance frost endurance etc.

Importance of Rootstock

Rootstocks can be used under the following conditions:

Impart salt tolerance: Saline injury leads to marginal leaf burning, premature leaf fall, stunted and weak growth and reduced yield. Many rootstocks like Kurakkan, Olour in mango, Rangpur lime, Cleopatra mandarin in citrus improve the salt tolerance limit of the scion cultivars.

Improved drought tolerance ability: In semi-arid tropics where evapo-transpiration is high, the plant suffers from soil moisture stress resulting in weak growth and poor crop production. Excess moisture stress leads to reduce fruit size with dull fruit colour. Rootstock provides vigorous root system to the scion, which is capable of developing the feeder roots and also reach farther from stem.

Improve nutrient uptake: Rootstocks vary in their capacity to absorb plant nutrients. The roots of rootstock increase the nutrient uptake efficiency.

Improving quantity and quality of produce: The rootstock modify the shoot vigour and improve the quality and yield.

Impart disease and pest tolerance: the rootstocks increase the tolerance of scion cultivars against diseases and pests.

Impart dwafrness: rootstock reduced vigour and size of scion cultivars and, hence vigorous cultivars also can be used for high density planting.

Citrus Rootstocks in India

- Rough lemon (*Citrus jambheri* Lush.) is the most widely used rootstock in the country for most of the citrus scions followed by Kharna Khatta (*Citrus karna*) being more popular in the Punjab and Uttar Pradesh.

- For mandarins Karna Khatta and Nasnaran, Rough lemon and Kharna Khatta for grapefruit are usually employed as rootstocks in the Punjab.

- Sweet lime is recommended for Mosambi under wet-subtropics with a high temperature. However, under north Indian condition, Karna khatta and Cleopatra mandarin are suitable rootstocks for mosambi.

- In western India, Jambheri is employed as a rootstock for Mosambi.

- Rangpur lime (*Citrus limonia*) proved to be the best for Nagpur santra due to its resistance to gummosis and root rot.

- In Assam, Rough lemon has been found to be the most promising rootstock commercially for Khasi mandarins, Valencia, Mosambi and Malta sweet oranges, lime, lemon etc., and pummelo for grapefruit.

- While Gajanimma proved superior for acid limes and lemons.

- Under Coorg conditions, for Coorg mandarin, both Rangpur lime and Kodakithuli orange (*Citrus reticulata* Blanco) have shown great promise as rootstocks.

- Gajanimma is recommended as a rootstock for Nepali oblong lemon.

- Sathgudi is recommended as stock for Sathgudi scion for disease resistance and good yields in Andhra Pradesh.

Characteristics of Some Citrus Rootstocks

Sour orange (*Citrus aurantium*)

Highly used worldwide, except in areas with tristeza, produces a tree with moderate to high vigor, size, and yield. Imparts excellent fruit quality characteristics to both sweet orange and grapefruit, although harvest may be delayed due to higher acid content. Complete resistance to phytophthora and high quality make this stock useful for grapefruit. Imparts cold hardiness to the scion. Most of the sour orange varieties are reported to be susceptible to quick decline. It is better tolerant to salts. It is well suited for heavy moist soil and withstands flooding for short periods better than others.

Rough lemon (C. *jambhiri)*

Deep rooted large trees; highly susceptible to *Phytophthora,* blight, low fruit quality; high early yields, sensitive to cold, good adaptability to light sandy soils, grapefruit trees on Rough lemon are susceptible while orange trees are tolerant to quick decline, Jambhiri of Kodur is highly susceptible to dry root caused by *Fusarium* species when Sathgudi orange is used as scion, fruit size is generally bigger.

Volkamer lemon (C. *volkameriana)*

Similar to rough lemon, but cold hardy, more tolerant to *Phytophtora parasitica.*

Rangpur lime (C. *limonia)*

High early yield, salt tolerance, fruit quality moderate, sensitive to *Phytophthora.* Highly resistant to tristeza and does well in heavy soils. More tolerant to salts than other rootstocks. Susceptible to exocortis. Trees on this stock are vigorous, precocious and prolific with quality produce.

Sweet orange (C. *sinensis)*

Tolerant to tristeza, fruit quality high, very susceptible to *Phytophthora,* moderate to high blight tolerance. It is a good rootstock for all the citrus cultivars, producing large, vigorous trees. Its greatest asset is that it is relatively tolerant to quick, decline and other diseases. It is adapted to well-drained, light to medium-loam soils. Not usually favoured for light soils because of its shallow and less widely spread root system.

Cleopatra mandarin *(Citrus reshni)*

Small fruit size, salt tolerance, cold tolerance, fruit quality high, slow growth in nursery; relatively tolerant to blight and high pH, having resistance to quick decline, gummosis and it is moderately cold tolerant. It thrives well on heavy soils and is better adapted to saline soils as compared with Rough lemon or sour orange.

Trifoliate orange *(Poncirus tnfoliata)*

Trifoliate orange is the hardiest and it imparts cold hardiness to the scion variety grafted on it. It has tendency to retard growth in spring by which it saves the scion from frosts. Trifoliate orange is not suitable for dry or calcareous soils. Perform well in sandy loams and clay alluvial soils with good moisture content and organic matter. It is susceptible to lime-induced chlorosis and is more sensitive to high soil sodium or potassium or combination of these, but tolerant to excess boron. Flying Dragon is a trifoliate mutant with serpentine stems, curved thorns, small leaves, and contorted habit can be used in pot culture since it truly dwarf. Fruit yield and quality are very good.

Citrange (C. *sinensis* x *P. trifoliata)*

'Carrizo', and 'Troyer', are the most common rootstocks commercially used. It is in general cold hardy and vigorous, fruit quality and yield is better on this rootstock. Susceptible to blight, exocortis, poor tolerance to salt and high pH. It is highly resistant to gummosis and tolerant to quick decline. Trees on Troyer are precocious and prolific, with large sized fruit of excellent quality. Troyer citrange and carrizo citrange are recommended for medium to shallow soils.

Swingle citrumelo (C. *paradisi* x *P. trifoliata)*

This hybrid is tolerant to tristeza and *Phytophthora parasitica* and is moderately tolerant to salts. It has shown satisfactory promise for sweet oranges and grapefruit. Valencia oranges on this stock are more cold hardy.

Alemow (*C. macropylla*)

High early yield, susceptible to tristeza, xyloporesis, fruit quality moderate to low, sensitive to cold and blight.

Mango Rootstock

In our country, mango is generally grown on non-descriptive type seedlings rootstocks.

However, recently it was reported that some polyembryininc genotypes can be used as a rootstocks for better yield, dwarf canopy and under environmental stresses.

Kurukkan: Salt tolerant, good rootstock for Amrapali

Olour: Salt tolerant, imparting dwarfness in Dushehari.

Vellaikoloban: Import darfness in Alphanso, allopolyploid having chromosome number 2n = 80.

Turpentine: salt tolerant, vigorous

Sucary and Sabre: Semi dwarfing rootstock

13-1: Dwarfing rootstock, having salt tolerance ability.

Pahutan: High yielder for Baneshan

Goa: High yielder for Neelam

Guava rootstocks

Seedling rootstock: Guava is generally propagated on seedlings type rootstocks. Produces variable yield and quality of same scion cultivars.

P. cattleianum : improves the yield of scion cultivars but not get commercial status.

P. friedrichsthalium : resistant to nematode, guava wilt, dwarfing but not used commercially.

P. cujavillis : Improves the quality of fruits of scion cultivars particularly ascorbic acid content but not used commercially.

P. Pumilum : Imports dwarfing, improved sugar content of scion cultivars.

Pusa Srijan: Released from Division of Fruits and Horticultural Technology, IARI, New Delhi, import dwarfness in Allhabad Safeda and showing tolerance against wilt under field condition.

Grape rootstock

In India, most commonly used rootstock in grapes is Dogridge. This was introduced from the University of California, Davis, USA, which was brought by Late Sh. Vasantrao Arve, a progressive grower from Sangli district of Maharashtra. There are two types of of Dogridge rootstocks were named as American Dogridge (Dogridge-A) and Bangalore Dogridge (Dogridge-B) respectively.

Drought Hardy Rootstocks

Vitis riparia grows in alluvial soils, climbs trees, and tolerates wetter areas. It has shallow roots, lower vigor compared to other species, and resistant to phylloxera. Originating in colder areas, the *V. riparia* is often more cold hardy than other species.

Apple rootstocks

Standard: Apple trees planted on standard rootstocks will produce large, full-sized trees that may grow more than 25 feet tall. They are very hardy and can be planted in a wide range of soils and climates. They are sturdy, long-lived (50

Rootstock	Parentage	Growth	Phylloxera resistance	Reaction to environmental stresses		
				Nematodes resistance	Drought tolerance	Salinity tolerance
5BB\	V. beriandieri x V. riperia	High	Excellent	Moderate	Low Variable	Low
110R	V. beriandieri x V. rupestris	High	God	Low	High	Moderate
1103P	V. beriandieri x V. rupestris	High	Good	Low	Moderate	Moderate
3309C	V. beriandieri x V. rupestris	Low	Good	Moderate	Low	Low
140R	V. beriandieri x V. rupestris	High	Excellent	Low	Excellent	Good
St. George	V. rupestris	High	Excellent	some	some	Moderate
SO4	V. riperia x V. beriandieri	High	Resistant	High	Low	-
99R	V. rupestris x V. beriandieri	High	Excellent	Moderate	Low	-
1613C	Solonis x Othello	Medium	Low	Excellent	Low	-
Dogridge	V. Champini	High	Low	Excellent	Low	
Salt Greek	V. Champini	High	Low	Excellent	Low	Good
1616C	V. riperia x	Low	good	Very susceptible	Very susceptible	

years or even longer), and productive. Following are vigorous rootstocks of apple.

MM.111

This tree is about the best in this class, although tree size is about 80% of standard. Though slightly more vigorous than the old M.2, the anchorage of MM. 111 is better and the tree seems adaptable to a wider range of soil conditions. Commercial stocks of MM.111 are virus- free and appear resistant to collar-rot.

Robusta 5

This is no longer recommended as a root or body stock in Ontario. It has a very short rest period and may break dormancy during a mild spell in winter. A subsequent freeze can result in sunscald, bark-split, cambium injury and even death of the tree. Where no frost injury occurs, the trees are difficult to manage, being more vigorous than standards. Robusta 5 was developed at Ottawa and should be considered as a rootstock only in the coldest districts where the winter is not broken by mild spells.

Semi-Dwarf: Trees planted using semi-dwarf rootstocks will reach a height of 15 to 20 feet. They should be planted at least 20 feet apart. They will produce about five bushels of full-size apples per year. They are not as long-lived as full-sized trees and have a life expectancy of about 20 - 25 years. They do not require staking. Semi-dwarf rootstocks of apple are mentioned here under.

M.7

A popular apple rootstock in North America because by chance it has good fireblight resistance and can be treated as a small MM106 in most respects. Released from the East Malling breeding program, this rootstock gives a tree a little larger than M.26 and a little smaller than MM.106. It is generally too vigorous for high density plantings. It is available in various virus reduces states such as M.7A, M.7 EMLA. It performs best on a good soil in a location protected from the wind, in a district with relatively mild winter temperatures. Bud high and plant deeply to improve anchorage and to reduce the strong tendency to produce root suckers. M.7 is a good producer of plants in the stoolbed,

V.4 (Vineland 4)

From the Vineland breeding program. Produces a tree similar in size to M.7. This rootstock is undergoing further testing at the University of Guelph, Simcoe Research Station, and at the time of print no schedule for commercial release is available.

MM.106

One of the most popular apple rootstocks, developed in collaboration by the East Malling and Merton research stations in the early 20th century, and derived from the Northern Spy apple variety. This rootstock makes a well-anchored tree whose size ranges from a large semi-dwarf to three-quarters of standard size. It is resistant to wooly aphid by susceptible to fireblight, Very susceptible to collar rot and Not cold-hardy.

Dwarf: Apple: Trees planted on dwarf rootstocks will grow 10 or 12 feet tall. They should be planted no closer than 12 feet apart. They will be less hardy than full-sized or semi-dwarf trees. Dwarf trees will live for about 15 to 20 years and will begin bearing fruit in two or three years. They will produce one or two bushels of full-size fruit a year, and because of their smaller size, most of the fruit can be picked without a ladder. Following are dwarf rootstocks of apple.

M.27

An extremely dwarfing rootstock released in 1971 from the East Malling breeding program, England. M.27 is probably too dwarfing to be useful in commercial orchards. It makes a tree about 20% the size of standard and half the size of on M.9 or smaller. As it is a slow, weak grower in the stoolbed, special techniques are needed to produce plants in quantities for commercial use. Winter hardiness is the same as M.9, and it does not produce root suckers or burr-knots. It is very precocious and is less susceptible to fire blight than M.9.

V.3 (Vineland 3)

This is a new dwarfing rootstock originating from the Vineland breeding program, Ontario. This rootstock is slightly less vigorous than M.9 EMLA but similar to the M.9 clones M.9 T337 and M.9 Flueren 56. Trees on V.3 have a similar propensity to form root suckers and appear to be as productive as the M.9 clones, but are more yield-efficient. Preliminary tests indicate it is moderately resistant to fire blight. This stock is presently being evaluated for winter hardiness, disease and insect resistance, incidence of burr-knots and root suckers, and susceptibility to collar-rot. The rootstock will be commercially available as early as 2001.

G.65 (Geneva 65)

This is a patented rootstock from the Cornell University breeding program, New York State. It is a very dwarfing stock producing a tree smaller than M.9. It is precocious and productive. It is resistant to fire blight and collar-rot and is moderately susceptible to woolly apple aphid. It has few burr-knots and few suckers.

M.9

One of the most widely-used rootstocks in commercial apple orchards, but not suitable for areas at risk of fireblight. Apple trees on M9 are very productive and come into bearing within 2-3 years of planting. It is the most dwarfing in commercial use. It produces a tree size approximately 25%-30% of full size with most cultivars. Trees on M.9 need to be supported throughout their lifetime. M.9 will not do well under poor drainage but it is tolerant of collar- rot and does well on heavier soils where drainage is adequate.

Vineland 1

Originated from the Vineland breeding program, Ontario. Tree size is comparable or slightly larger than M.26. Yield efficiency and fruit size are equal to or greater than M.26. Propensity to sucker is equal to M.26. Unlike M.26, V.1 appears to be highly resistance to fire blight.

Budagovsky 9

It Is a dwarfing rootstock bred in the Soviet Union. Its leaves are red. Tree size falls between M.26 and M.9 EMLA in vigour. It is a precocious rootstock with high yield efficiency. It requires support. Bud.9 is resistant to collar-rot and susceptible to fire blight and woolly apple aphid. Bud.9 has excellent winter hardiness much greater than M.9. It produces few suckers or burr-knots. It is a promising dwarfing winter-hardy rootstock.

Ottawa 3

This is the most dwarfing rootstock to come out of the cold-hardy breeding program at Ottawa. It appears more dwarfing than M.26 but more vigorous than M.9. Ottawa 3 roots sparsely in the stoolbed but can be grown from root cuttings or tissue culture. It is more cold-hardy than M.26 or M.9 and is resistant to collar-rot but susceptible to fire blight and woolly apple aphid. It does not produce burr-knots or root suckers. It is precocious and requires support while young.

M.26

A dwarfing rootstock introduced from East Malling in 1959. This rootstock has been quite popular in the last 10 years. M. 26 is recommended in all apple districts of Ontario, but on a trial basis only in colder districts. M.26 is reported to be the most hardy of the Malling series rootstocks. The tree is about 40% of standard size, being larger and sturdier than M.9 but smaller than MM.106. It will do well on intermediate or heavier-textured soils if drainage is adequate. While moderately resistant to collar-rot, M.26 will not perform satisfactorily on poorly-drained sites. It is resistance to collar-rot and apparently susceptible to fire blight.

Geneva 3

A patented rootstock released in 1994 from the Cornell University breeding program, New York State. This is a Robusta 5 X M.9 cross 60% - 65% the vigour of seedling, similar to M.26. This rootstock appears to be more productive than M.7 but similar to M.26. It is bred for its resistance to fire blight.

Vineland 2

From the Vineland breeding program. Tree size is 20% larger than M.26 based on trunk cross-sectional area and tends to be wider while similar in height to M.26. Productivity is equal to or slightly greater than M.26. Premilinary fire blight testing of V.2 indicate it is moderate-highly resistant - comparable with M.7. Tree survival of V.2 is comparable or better than M.26, especially in colder climates. The rootstock will become commercially available as early as 2001.

Spacing of apple cultivars of average vigour on certain leading rootstocks

Rootstock	Meters		
	Low Density	Medium Density	High Density
M.9	----	2.5 x 5.0	1.5 x 3.5
M.26	4.0 x 6.0	3.0 x 5.5	2.5 x 4.5
M.7	5.5 x 8.0	4.5 x 6.5	3 x 4.5
MM.106	6.0 x 8.5	5.0 x 7.5	3.5 x 6.0
MM.111	6.5 x 9.0	5.5 x 8.0	----
Vigorous Rootstocks	7.5 x 10.0	6.0 x 8.5	----

Peach Rootstocks

Tolerance of Peach Rootstocks to biotic and abiotic Stresses.

Rootstock	Root knot Nematodes	Calcareous Soil	Water logging	Cold Hardiness
Lovell	S	MS	S	Moderate
Halford	S	-	S	Moderate
Nemaguard	R	VS	S	Poor-Fair
Nemared	R	S	S	Fair?
Guardian	R	MS	S	Moderate
Flordaguard	R	VS	-	Poor
Titan Hybrids	R	R	VS	Fair-Good
Hansen	R	R	VS	Fair-Good

VS = very susceptible; S = susceptible; MS = moderately susceptible; R = resistant.

Pear Rootstocks

Rootstock	Vigour	Distance	Forms of tree	Remarks
Quince A	Medium	2.5 x 2.5 M	Bush, cordon, Small fan or espalier	Produced the smallest trees, best for small garden, permanent or other stakes are required, comes into bearing after 3-4 years
Quince B	Large	3.5 x 3.5 m	Bush, large cordon, Large fan or espalier	The best general purpose pear rootstock, comes into bearing after 4-5 years.

Plum Rootstocks

St. Julien

The most versatile rootstock for plums, gages and damsons is St Julien. St. Julien is a member of the species Prunus insititia, which also includes damsons and mirabelles. St. Julien rootstocks produce a tree which is substantially smaller than plum trees grown on their own roots. It is compatible with almost all plums and gages. In fact it is also widely used for peaches, nectarines, and apricots, which are very closely related to plums. Plum trees grown on St. Julien rootstocks tend to come into bearing after 3-4 years. St. Julien can be considered roughly equivalent to the apple MM106 rootstock in the size of tree it produces.

Pixy

A further selection of St. Julien called Pixy is now also used as rootstock. This produces a somewhat smaller tree than St Julien, and is also slightly more precocious - the tree will bear fruit about a year earlier than the same variety grafted on St. Julien. Trees grown on this rootstock will need staking for the first 4-5 years, and prefer from better soil conditions and watering than trees on the St. Julien rootstock. Pixy can be considered roughly equivalent to the apple M26 rootstock in the size of tree it produces.

4

Layout of Modern Nursery

What is Nursery?

Nursery is a place, where seedlings, samplings, trees and other plant material are grown and maintained until they are planted in permanent place. Setting up of a horticultural nursery is a long term venture and needs proper planning for good return.

Establishment of Nursery

To be a viable venture, the nursery should be established in such an area where cultivation of fruit crops is on sizeable area and there is need for a nursery, having demand for saplings. In such area/region, following considerations need to be observed for selecting an appropriate location.

1. Nursery should be established at central place.

2. Nursery soil should be deep, fertile, well drained and free from pathogens.

3. It should be well connected with roads/transport media.

4. The area should be well protected.

5. Soil and micro-climatic conditions should be appropriate.

6. Availability of irrigation and power supply is ensured.

7. Availability of skilled personnel s is ensured.

8. Availability of mother stock and root stocks be ascertained

Components of Modern Nursery

A number of structures may be necessary for raising a nursery. To begin with, the following structures need to be constructed:

1. Fencing

Proper fencing should be done to protect a nursery particularly from stray animals. For a model of 0.5 acre area, an amount of Rs.30000 has been considered as the total cost for erecting a fencing around the boundary.

2. Progeny block (bud wood source tree)

In progeny block, true to type mother plant are maintained in the nursery. Suitable plant types with existing superior cultivars/varieties should be collected and maintained in the progeny block. The mother plants should be true to type, healthy, heavy bearer with standard quality fruit having tolerance to biotic and abiotic stresses. Care should be taken to label the plants properly. The bud wood source trees are maintained by adopting appropriate sanitary procedures. One is to know the diseases, which are to be taken care of. Some diseases are only transmissible by graft and caused by virus, some by contaminated pruning tools, some transmissible by seeds. In addition, infectious diseases may be caused by surface pathogens, which are not graft transmissible but may contaminate nurseries and inflict severe economic losses.

3. Rootstocks and seed gardens

After establishment of scion bank, next priority should be given for the establishment of the rootstocks. Seed propagation is the most usual way for mass production of rootstocks. There should be seed gardens comprising of the seed source trees to produce rootstock seeds. While establishing seed gardens, the rootstocks should be planted into separate blocks with windbreak and shelter belts to avoid cross-pollination and keep plant trueness to type. The plant-to-plant spacing for seed gardens should be kept close. The basic information should be taken into account for planning of seed gardens, which must be properly maintained to avoid diseases caused by phytophthora, nematode, bacteria and virus.

4. Growing structures

There should be provision of modern propagation structure like greenhouse/polyhouse, mist chamber etc. these structure provide optimum growing conditions for seed germination, rooting of cutting, hardening of seedlings. Now a days green house has become a prerequisite of Hi-Tech nursery. The greenhouse helps in providing additional carbon dioxide to the plants to enhance their rate of photosynthesis. This can be achieved by enclosing the plants in a box-like structure made out of bamboo and colourless transparent plastic, with a lid at the top.

The polyhouse of 9 m x 4 m dimension with 90 cm, brick wall, 3.6 m tall rhombus netting with expanded metal and polythene roof supported by local materials like bamboo, wood and planks, may be constructed. The cost estimated for such a house is approximately Rs.300.00 Shade nets are useful not only for reducing heat injury to young plants, but their use also reduces transpiration. Shade nets are available in different colours and densities.

5. High Humidity Chamber

This technique resolves the common problem of grafts or cuttings dying due to desiccation (drying up) when planted in the soil for rooting, by ensuring a humid atmosphere around the cuttings, thus preventing excessive evaporation. The cuttings/grafts are planted on a sand bed, enclosed on all the sides by a dome made of GI wire and covered with a transparent, colourless plastic film. The sand is watered to field capacity, and the plastic film traps evaporation inside the chamber creating a highly humid atmosphere. The dome must be shaded, since direct sunlight will heat up the internal atmosphere of the dome, killing the plants.

6. Use of Supplementary Light

Several plants go into winter dormancy when the day length gets short. Additional light from tube lights, given after sunset, creates long-day condition that prevents the plants from going into winter dormancy. Light, given at the end of the day (photo-period), also encourages growth of green leaves; they grow tall without developing lateral branches. On the other hand, if they are exposed to fluorescent light from tube-lights laid on the ground, they develop side branches and show a bushy habit. This occurs due to phytochrome (a pigment in all green plants) which detects different kinds of light, leading to an appropriate growth response. Exposure to light is necessary only for about half an hour immediately after the end of the photo-period to get the desired results. Scientific use of supplementary light substantially enhances the growth of plants.

7. Drought Hardening

Plants that are raised under high atmospheric humidity and shade, often die due to transplanting shock when shifted to the fields. To prevent this, the plants are hardened by allowing external dry air to enter the chamber gradually. This is achieved by lifting the plastic film on two opposite sides of the high humidity chamber to some extent to create small openings in the chamber. The openings are widened every day, in such a way that the entire film can be removed after about 8 days. In this way, the plants get adapted to dry air gradually. Such plants can be transplanted in the fields but as precautionary measure it is advisable to transplant on a rainy or cloudy day.

After having established the above infrastructure, the nursery establishment and planning involve division of the nursery into three different units, viz., propagation unit, production unit, sale unit and packaging unit.

1. Propagation unit

Propagation unit is the major unit of nursery work and includes:

(i)   Actual propagation structures such as green house, hot beds, cold frames and mist house.

(ii)   Service structures such as head house.

(iii)   The alley house connecting to the hot beds and cold frames with head house. It provide a passage for the transport of plants, propagation media, soil and fertilizers from head house to propagation structures, and must be sufficient wide to permit easy and quick movement.

(i) Primary nursery (Seed beds)

Seed beds near to water source and to office so that they can be kept under vigilant control. The raised seedbeds of 6-8 cm height, 1 meter width and of convenient length, free from stones should be prepared with upper 2.5-5cm of the bed filled with sand. Soil can be prepared to fine tilth, add sufficient quantity of rotted FYM, vermin-manure or pig manure at least 10-15 days earlier of seed sowing. The bed may be treated with 1% Bordeaux or 0.1% Bavistin before sowing of seeds.

(ii) Nursery beds

Seedlings from seed beds are removed and transplanted in the nursery beds. Nursery beds should be located in an open area near to water source. Nursery beds should be prepared by adding sufficient organic manures and fertilizers. Nursery beds should be divided into section as per crop and varieties. The nursery beds should be laid out in such a way that there is an access to all the beds through roads or paths.

(iii) Pot yard

The pot yard should be in shade because the tender plants require shade as compared to hardy plants. This section should be near to water source. Trenches can be provided for keeping potted plants closely packed together.

2. Production Unit

The object of this unit is to rear the new plants from seedling to marketable stage. This unit is divided into different blocks, each block being meant for only one kind of plant or species. This helps on sorting of plants, easy record keeping and doing the operation as per the need of a plant species.

3. Packaging Unit

The packing yard is used for packing the plants before sale or dispatch to out stations. The yard can be combined with working shed. It is near to sale counter.

4. Sale Unit

The objective of the sale unit is to market the nursery plants effectively. The design and layout of this unit should be attractive to the customers. This should be usually located on a well travelled way and may be by the side of production unit. The sale unit is usually divided into three different parts such as display unit, sale and packing, and parking unit.

In the display or show unit, plants are displayed in such way that, it attracts more customers. Ornamental plants are used for effective arrangement using combinations of trees, shrubs and various topiary works. The sale area represents actual sale counter and handing over of the plants. Parking area is also planned in sale unit for easy parking of vehicles of customers and others. The walk ways should be generally wide enough to permit the use of wheel barrow, and benches should be narrow so that the centre can be reached from either side.

5

Packaging of Nursery Plants

The packing methods practiced by nurserymen for transport of nursery plants to distant places in India are neither uniform nor efficient. In different parts of the country, various practices are followed which are claimed to have been evolved by long experience of the local nurserymen. They vary according to nature of the plant, the season and the cheapest packing material available at the time of packing.

In many parts of the country, citrus nurserymen uproot plants, tie them in large bundles (with or without soil), keeping the plant tops exposed to air and roots tightly packed or tied in the gunny cloth. The pomegranate and guava plants prepared by air layering are packed as such if they are not already potted in small pots.

Many nurserymen pack the rooted or newly lifted plants in a ball of sticky clay. The ball of earth is then covered with grass and tied as firmly as possible, thus securing the roots against drying. A number of plants, thus prepared, are filled closely in light locally knit bamboo baskets. The baskets are covered thinly with coconut leaves as protection against injury during railway shipment. Mango grafts in pots are packed in light baskets with some cushion of grass to reduce the impact while loading and unloading the parcels on their long journey by rail and road.

Some nursery men use large baskets, woven out of coconut leaves with limp walls or sides. Plants such as coconut, chikoo, guava, mango are packed tightly in these baskets and shipped by truck or rail. This packing is done irrespective of plant containers such as pot or a ball of earth, with or without grass covering. The plants packed in all such fashions are usually damaged and some of them do not survive, unless care is taken during or after transplanting them in the permanent places. For better success, it should be seen that the roots of grown up plants do not get irreparable injury while removing from the bed or the pot. During transport, loss of water by transpiration should be kept at minimum, while the plants are awaiting dispatch or delivery at the railway platform, precaution should be taken against pilferage or damage by animals.

Efficient Packing

(i)   If plants are watered lightly, before removing from the bed or pots, they can be taken with reduction in injury to the roots.

(ii)   A small ball of earth (10 cm dia) is retained around the roots, trimming those which are excessively long.

(iii) The plant is packed in a small piece of gunny cloth (1 to ½ sq feet). The four ends of this cloth are brought over the ball (earth covering the roots) and tied fairly tight. Thus the moist ball of earth is secured round the roots and there is no chance of its breaking loose during the journey.

(iv)   In case of mango and sapota, where it is desirable to retain the earthen pot, the gunny cloth is wrapped around the container and tied firmly. During transit, if the pot is broken, the gunny cloth keeps the broken shreds in place and damage to the soil and plant roots is minimized.

(iv)   A round, strongly woven bamboo basket is selected which is reinforced with iron or strong bamboo strips at the base. Some grass, dust or any other light material such as paper shavings or leaves is spread inside the basket to act as cushion. The plant roots are soaked in water and packed in the basket. The tops are tied to prevent their stems from shaking, jagging or rubbing together. Three strong bamboo sticks of about 60-75 cm long are inserted firmly into woven material of the basket and tied at the top. These form the frame work for the small gunny cloth tent which covers the basket and most of the plants. The gunny bag cloth cover is sewn up with the basket and along the joint. The address label for the parcel is covered with polythene bag to preserve it against water and weather. This method is comparatively expensive and tedious than other packing practices prevailing at present. But more survival and safety of plants is assured.

Thicker gauge poly bags or wrappers are also suggested to pack the ball or earth (soil) at the roots. This involves less labour and keeps packing cost down. To pack the plants for air transport, the size and weight of the package is limited. The plants are to be in limited number selected and not more than 30 cm in length, including the roots. Branches or leaves are trimmed off or defoliated, without causing injury. The soil is washed off the roots of the plant by shaking its lower part in water. Hormone powder is sprinkled on the roots and damp moss is tied, in a ball to the roots. This is encased in a polythene bag.

A light hard cad board box is suggested to pack the plants. Damp moss is spread over the lining of the water proof paper in the basket. The plants are laid over this cushioning of moss and fixed firmly with strings to prevent shaking of plants. The basket or box is closed in a piece of muslin cloth. The address label is firmly sewed on it.

While replanting the plants received by rail, it is recommended that the plants be planted without removing/disturbing the ball of earth. It is also suggested that moss is removed and roots are soaked in water for few minutes before transplanting.

Usually the plants shed their leaves during a long journey when their roots are packed in moss. But they sprout again after repotting or transplanting them at the destination.

The packing method suggested above may be changed to suit individual plant requirement, while adhering to the basic principles enunciated here.

6

Preparation and Application of PGR in Fruit Crops

Plant Growth Regulator (PGR) are diverse group of organic compound other than nutrients, produced artificially or by plants which in low concentrations promote, inhibit or modify the physiological process in plant. Among the PGRs produced by plants are auxins (indole-3-acetic acid and its derivatives), gibberellins, cytokinins, abscisic acid and ethylene.

Application of certain plant growth regulators promotes the process of seed germination, and development of roots in cuttings and layers. Various growth regulators have been used with striking success. Of these auxins such as indole-3-acetic acid (IAA), indolebutyric acid (IBA) and naphthalene acetic acid (NAA) have the greatest effect on root formation in cuttings and layers. These chemicals speed up the healing of the wound, induce more roots and their development. They are now used universally in the propagation of many fruit plants. The growth regulators are effective in minute concentration and are usually applied in dilute solutions and dust formulations or rarely in grease bases such as lanolin. Plant growth regulators are generally calculated in the form of parts per million (ppm) or microgram per gram (mg/g).

Materials Required

1. IBA or IAA or NAA	5. Camel hair brush
2. Talc Powder	6. Alcohol (95%)
3. Lanolin	7. Glass rod
4. Measuring cylinders	8. Petri dish

Formulation of Growth Regulators

1. Solution form

The solution form is normally used for rooting of cuttings. There are two main ways of solution application: the dilute solution method and the concentrated solution method.

(a) **Dilute solution or prolong dip method**

This method depends upon the slow uptake of comparatively dilute water solutions over periods varying from 8-24 hours. In practice, it is convenient to place the cuttings in the solution at the end of the day work and to plant them next morning, some 16 hr later. The cuttings are allowed to stand in the solution with about 2.5 cm of their based immersed. Basal leaves, are first removed. It should be ensured that all cuttings remain in the solution throughout the treatment period. After that, they are rinsed in plain water and planted. Low concentrations is used for easy to root plants while higher concentration for difficult to root species. Prolong dip method is used in combination with hormones, vitamins, sugars and nitrogenous compounds for encouraging adventitious rooting in difficult to root species.

(b) **Concentrated solution method or quick dip method**

The customary procedure is to momentarily dip the basal 0.5-1.0 cm portion in growth regulator solution and plant the cuttings immediately. Dipping for a short time (say 5 seconds) in a suitably high concentration is probably the most convenient practical way of treating large number of cuttings. For this purpose, generally hormonal concentration varying from 500 to 10, 000 ppm is used. After treating the cuttings it should be immediately planted in rooting media. This method can be employed for treating large number of cuttings in small quantity of solution

2. Paste Form

Lanonin is a wool fat, semi solid, greenish yellow in colour, which can be made into liquid form, just by gentle heating. Lanolin paste is particularly convenient for use in air layering though it is widely used for cuttings.

3. Dust form

Certain growth regulators and their commercial formulations are available in powder or talc form e.g. Seradix A and Seradix B, which may directly be used for treating the cuttings. The usual method of application is to dip the bases of the cuttings, a number at a time, into the dust so that the lower 2.5 cm or less base is covered. If the dust does not adhere to the cutting, the bases of the cuttings may be wetted with water, before dipping into the dust. It is better to place a suitable quantity of the dust in a shallow dish sufficient only for the task on hand. Once used, the surplus material should not be returned to the stock. After dipping the cuttings into the dust, they should be lightly tapped on the side of the vessel so that the excess powder falls back immediately. They should not be pushed into the medium, as this removes most of the dust, but should be placed in a small trench or dibbled hole.

Aerosol form

Growth regulators are used in aerosols form in the greenhouses for rooting in soft wood and herabaceous cuttings. Mother plants may also be sprayed with hormonal solution prior to obtaining the cuttings. The concentration of 25 to 100 ppm may be kept for such sprays which are performed 30 to 40 days prior to taking cuttings from such plants. This process enhance the rooting in cuttings.

Preparation of Growth Regulators

1. Solution form

Plant growth regulator are applied at very low concentration and in exact quantity. Higher or lower concentration of growth regulators may result into negative effect particularly PGR s like 2,4-D. Growth regulators being less soluble in water it should be first dissolved in small quantity of a solvent alcohol, potassium hydroxide etc. solvent and then finally dissolved in water for making the desired solution. A list of solvents for different commonly used growth regulators are given below.

Table Solvents for dissolving plant growth regulators

Hormone group	Name of growth regulator	Solvent (s)
Auxins	Indole Acetic Acid (IAA), Indole Butyric Acid (IBA), 4 CPA, NAA 2, 4-D and 2, 4,5-T	Ethyl alcohol, Methanol, Potassium hydroxide, Sodium hydroxide Water
Gibberellins	Gibberellic Acid (GA3), 6-Benzyl Adenine, Benzyl Amino purine, Kinetin2 ip	Etanol or Methanol0.1 N HCL
	CPPU	Water
Ethylene	Ethrel	Water
	Cycocel (CCU), Alar (SADH) Malic Hydrazide (MH), Dormes	Water

Example: For making a solution of 5,000 ppm IBA in 500 ml volumetric flask

Normally growth regulators are expressed in ppm, which means parts per million. This is equivalent to 1mg/litre or 1µg/ml or 1mg/kg. For making the solution following formula may be used.

$$\text{Ppm} = \frac{\text{Plant growth regulator (mg)}}{\text{Required volume (ml)}} \times 100 \text{ or}$$

$$\text{Plant growth regulator (mg)} = \frac{\text{Required volume (ml)} \times \text{desired concentration (ppm)}}{1000}$$

Therefore, for making 5,000 ppm in 500 ml water, the following quantity of growth hormone will be required:

$$\text{Plant growth regulator (mg)} = \frac{5000 \times 5000}{1,000} = 2,500 \text{ mg IBA}$$

Hence, the 2,500 mg IBA will be first dissoloved in 50 ml ethyl alcohol. Then it will be poured drop by drop in 450 ml of water and mixed by shaking. If some precipitations are seen in the solution, adding few drops of sodium hydroxide or potassium hydroxide will make a clear solution. The solution is ready for use. This solution can be stored in refrigerator (4°C) for 7-10 days until use.

2. Dust form

To prepare a dust containing 10, 000 ppm of growth regulator, dissolve 1g regulator in 40ml alcohol and stir this into 100g of pharmaceutical talc to form a smooth paste. This should be done in a dark room away form strong light. The paste should be stirred to avoid drying, until it becomes a fine dry powder. This dust remains active for six months or more if stored in a closed opaque container in a refrigerator. Before using, dilute this stock with talc powder.

3. Lanolin paste

It is prepared by stirring the growth regulator into the molten lanolin and then allowing it to cool.

To make a paste containing 5, 000 ppm of growth regulator, melt 200 g lanolin and stir into this 1 g of required growth regulator. This paste keeps longer if stored in a well opaque glass vessel, in a refrigerator.

7

Planning and Layout of Orchard

There are several planting plans or systems which can be adopted for planting an orchard. The main principle for the layout of orchard is:

(i)   It should accommodate maximum number of plant per unit area.

(ii)   Provide adequate space for the development of the tree.

(iii) Ensure convenience in orchard operations.

The various lay out systems used are the following

1. Vertical Row Planting Pattern

(i) Square

In this system of planting the plants are planted in straight rows running at right angle. The distance between plants and between rows being same four plants make a square. This method is normally preferred by the orchardist because short lived filler trees can be planted in centre. This is also easy to layout on ground and cultural operations will be rendered easy. Better watching is possible and cultivation and irrigation are possible in two directions.

(ii) Rectangular System

In this system, trees are planted on each corner of a rectangle. As the distance between any two rows is more than the distance between any two trees in a row, there is no equal distribution of space per tree. The wider alley spaces available between rows trees permit easy intercultural operations and even the use of mechanical operations.

(iii) Cluster system

In this system, trees are planted on each corner of a square forming a cluster and each cluster is set apart at double the distance of trees planted in a cluster. Although there is no equal distribution of space per tree, the wider alley space around each cluster permits easy cultural operation. It accommodates nearly twice the population of square system.

Alternate Row Planting Pattern

(i) Triangular System

This system is similar to the square system of planting except that in every alternate row the plants are planted in midway of two plants of the previous row. Thus, tree plants make a triangle where only two arms are of equal length.

(ii) Quincunx or Filler

This system is essentially the square system except for an additional tree in the centre of each square. Thus the number of trees are nearly double than the square system, but does not provide equal spacing. Center (filler) trees may be short lived. This is difficult layout on ground and can be adopted when spacing for permanent tree is more than 30 feet. This is not satisfactory as a permanent plant but is satisfactory for putting temporary trees in the centre of squares. Filler should be removed after a few years when main trees come to bearing.

(iii) Hexagonal or Septule

This system is also known as equilateral triangle system of planting. This system is also called septule because seventh tree is put in the centre of the hexagon. The plant in this system are planted at the corners of the equilateral triangle with one tree in the centre. Thus, six trees make a hexagon with an additional tree in the centre of the hexagon. The perpendicular distance between any two adjacent rows is equal to the product of 0.866 x the distance between any two trees. As the perpendicular distance between any two rows is less than unity this system allows 15% more plants than the square system. The limitation of this system is that it is difficult to lay out and the inter cultivation is not so easily done as in the square system.

(iv) Contour

This is necessary for rolling topography. Trees can be planted on terraces or along contours. Tree position can be decided only on spot. Contour planting is good for shallow soils where terracing will expose rocky or poorer sub soil. The contour line is so designed and graded in such a way that the flow of water in the irrigation channel becomes slow and thus finds time to penetrate into the soil without causing erosion. Trees will not equidistant in this method. Irrigation and cultivation can be done along tree row only. Where land is very steep terracing should be done across a sloping side of the hill, lying along the contours. Terraced fields rise in steps one above the other and help to bring more area into productive use and also to prevent soil erosion. In South India, tea is planted in contours either in single hedge system or in double hedge system. Double hedge contour planting system accommodates nearly 22% higher population than single hedge system. Number of plant population that can be accommodated in this system is

Plant population = $\frac{N \times Unit\ area}{D\ (y+z)}$

Where

N= Number of Hedges   D= Distance between plants

Y= Distance between hedges   Z= Vertical distance between row

The total number of trees per hectare for various important horticultural crops under (a) Square (b) Hexagonal and (c) Triangular system of planting are given below:

Crop	Planting distance (m)	No. of trees per hectare		
		Square system	Hexagonal System	Triangular System
Mango	10 x 10	100	115	89
Sapota	8 x 8	156	178	139
Lime	5 x 5	400	461	357
Coconut	7.5 x 7.5	177	205	159

Principles to Decide Planting Distance

(i) When fully grown, the fringes of trees should touch each other without the branches interlocking.

(ii) The roots will spread over a much larger area than top, hence more room should be allowed for roots to feed without competition.

(iii) Planting distance to should be based upon (i) Type of Fruit (ii)Varietal factor (iii) Rainfall (iv) Soil type and soil fertility (v) Root stock (vi) Pruning and training (vii) Irrigation system.

Selection of Plants

Careful selection of plants is necessary. The plants should be of good parentage, should have been propagated on the right root stock, should be free from pests and diseases and should have a healthy bark.

The age of trees at planning is also important. Older plants are much easily damaged in transit and establish poorly. They take long to revive. Old plants are not therefore preferable. It is better to plant within one year of grafting or budding and one year old plants which are 2 to 3 feet in height and should be selected.

Planting Procedure

For laying out the orchard by any system of planting, it is necessary that the first row is straight and running parallel to the boundary of the field. The lay out should be done in accordance to pythogorus theorm. The first row is drawn at half the distance allowed between the rows and within plants. For making first row, fix bamboos at an interval of 3-4 meters. Stand at one end of the field and see that all the bamboos are in one straight line, if not, adjust bamboos by shifting either side. Continue the process till all the bamboos fall in one straight line.

For hexagonal lay out use two chains. After putting the pegs in the first row, put chain in two pegs. Hold the chain at desired distance, now make an arc form both pegs. Put a peg where both bisect each other. This process is repeated till the entire field is covered.

8

Training and Pruning

Training and pruning are an important activity in fruit crops to have better frame work and optimum fruiting area. Training refers to giving a desired shape to the plants by tying or staking or supporting over a structure and or selective pruning for a good strong frame work. Pruning refers to cutting of certain portion of plants for maintenance of fruitfulness and quality besides vigour of the trees or vines. Pruning affects the functions of the plants and assists in better fruiting and in getting more quality fruits. It is one of the most crucial operations and require some scientific knowledge regarding bearing behavior of the plants.

Objectives of Training

The major objectives of training are

i. To give a strong frame work to the tree for supporting good cropping.

ii. Provide good exposure of light and air to branches and leaves.

iii. To maintain tree growth in such a way that that various cultural operations, such as spraying, annual pruning, harvesting etc can be done at the lowest cost.

iv. To protect the tree from sun burn and damage.

v. To secure a balanced distribution of fruit bearing parts on the main limbs of the tree.

vi. Maintain the vitality of trees over a long period of time.

Methods of Training

There are three most commonly used training methods are followed in fruit crops based on the growth habit of the fruit tree. These are

1. Open Centre

In this system of training, the main stem is allowed to grow only upto a certain height, thereafter it (leader or main stem) is headed back to encourage lateral branching (scaffold branches). This system is also known as **Vase-shaped**

system. This system allows better distribution of sunshine and to reach it to branches of trees and also facilitate cultural operations like spraying, thinning, harvesting etc.

2. Central Leader

In this system of training, main stem (leader) is not headed back and is allowed to grow I its natural ways extending from surface level to the top of the tree. This results in robust close centre and tall tree and branches are more fruitful near the top as compared to lower branches. This system of training is also known as closed centered one.

3. Modified leader

It is intermediate between the open centre and central leader training system.In this system main stem is allowed to grow unhampered for the first four or five years, thereafter it is headed back and lateral branches are allowed to grow as in the open centre system. Modified leader system produces fairly strong and moderately spreading trees.

Trees are trained to different forms with or without the support of certain structures.

Head System

It is mostly used for spur bearing grape cultivars. In this system, vines are trained like a small bush. Vines are allowed to, grow upto 1.2 meters, and then headed back to produce laterals. Four laterals- one in each direction is allowed to grow and rest are thinned out. In next dormant season, these laterals are cut back to 2 buds and further two arms of 20-30 cm are allowed on each secondary arms. After 3-4 years these vines will give a dwarf bush like appearance and requires no staking. Other training systems which requires no staking are Palmette, Spindle bush, Dwarf pyramid and Head and spread systems.

Bower system

It is also called as „Pandal or „Arbour or „Pergola system. It is generally practiced in grapes and other cucurbitaceous vegetables like snake gourd, ribbed gourd, bitter gourd etc. In this system, the vines are spread over a criss cross net work of wires, usually at 2.1 to 2.4m above ground, supported by concrete or stone pillars or live support like **Commiphera sp**. The vine is allowed to grow single shoot till it reaches the wire net and is usually supported by bamboo sticks tied with jute thread. When the vine reaches the wires, its growing point is pinched off to facilitate the production of side shoots.

Cordon and Espalier system

Plants are trained to grow flat on trellis or on horizontal wires by training the branches perpendicularly to the main stem on both the sides, and trained horizontally on the wires. Plants trained in this systems are called **'espaliers'**. An espalier with one shoot or two shoots growing in opposite or parallel directions are called a **'cordon'**.

Kniffin system

In this system, two trellis of wire are strongly supported by vertical posts. The vines such as grape when trained in this system has four canes one along each wire and the bearing shoot hangs freely with no tying being necessary.

Overhead trellis or Telephone system

This system consists of 3 or 4 wires usually kept at m45-60 cm apart fixed to the cross-angle arms supported by vertical pillars or posts. Vines are allowed to grow upto a height of 1.5 to 2.0 m and then trained on this system. Moderately vigorous cultivars with apical dominance are best trained on such system.

Tatura trellis

In this system, trees are trained to a multi-layered wire trellis. The trellis is V-shaped, supported by two long, stout poles embedded into the soil angles of 60° from the horizontal. Five wires at 60cm intervals are fastened to these poles. This system, is being now followed for pome fruits, nut fruits and grapes. The trees are grown as double leader. Trees with each leader inclined at an angle of 60° from the horizontal.

Pruning

Commonly, trees are pruned annually in two ways. A few shoots or branches that are considered undesirable are removed entirely without leaving any stub. This operation is known as „thinning out. The other method which involves removal of terminal portion of the shoots, branches or limb, leaving its basal portion intact, is called „heading back . Thinning out involving large limbs as in old and diseased trees is called „bulk pruning . Pruning is done with the following specific objectives.

i) To remove surplus branches,

ii) To open the trees so that the fruits will colour more satisfactorily

iii) To train it to some desired form

iv) To remove the dead and diseased limbs

v)   To remove the water sprouts and

vi)   To improve fruiting wood and to regulate production of floral buds.

Season of Pruning

Under South Indian conditions, old non bearing mango trees are pruned during August – September. The pome fruits such as apple, plum, pears and peaches are pruned every year in December - January. Jasmines are pruned to 45cm height from the ground level during the last week of November.

SPECIAL PRUNING TECHNIQUES

1. Root Pruning

A circular trench of 45cm away from the stem is dug out annually and the roots are cut- off every year with a sharp knife. After pruning, the trench is filled with manures liberally.

The tree is thus fed and watered artificially in a restricted area. Each year prune 4 to 5 cm of the stumps of the previous year growth. This helps to increase the production of mass fibrous roots, dwarf the trees and bears abundantly. This practices is not advocated every year to the fruit trees.

2. Ringing

It is one of the known practices to increase fruit bud formation in certain fruit crops. The operation consists of removal of a complete ring of bark from a branch or the trunk. Ringing interrupts the downward passage of carbohydrates through the phloem and thus causes them to accumulate in the part of the tree above the ring. Ringing is practiced on Mango to force flowering in over vegetative trees which do not normally bear a satisfactory crop. This practice cannot be recommended for all fruit crops and it is found beneficial in promoting fruit set in certain vigorously growing grape varieties and they often result in large size fruits.

3. Notching

Notching is a partial ringing of a branch above a dormant lateral bud. Eg. Fig, apple etc.

4. Smudging

It refers to the practice of smoking the trees like mango, commonly employed in Philippines to produce off-season crop. Smudging of Mango trees in India has not been found to induce early blossom.

5. Bending

Bending of branches is widely practiced in the Deccan for increasing fruit production in guava, especially in the erect growing varieties.

6. Coppicing

This refers to the practice of complete removal of the trunk in trees like Eucalyptus and Cinchona leaving 30-35cm stump alone.

The coppiced stump starts producing many vigorous shoots in about 6 months time. Only 2-3 shoots are retained per stump and the rest ones are completely thinned out. These left out shoots attain coppicing stage in about 10 years depending upon the locations and other factors.

7. Pollarding

This refers to the practice of removing the growing point in shade trees especially in silver oak in order to encourage side branches.

8. Lopping

This refers to the practice of reducing the canopy cover in shade trees in order to permit more light.

9. Pinching

Carnation, chrysanthemum to reduce the plant height and to promote auxillary branching.

10. Disbudding

The practice of removing unwanted flower buds in a cluster so as to encourage the remaining buds to develop into a large, showy, quality bloom is called disbudding. This practices is commonly done in cut flowers like carnation, chrysanthemum, dahlia, marigold and zinnia etc.

Thinning

Fruiting is an exhaustive process to the tree especially if the crop is heavy.

The other objectives of fruit thinning are the following :

1. To increase the annual yield of marketable fruit.
2. To improve the fruit size.
3. To improve the colour of the fruit.
4. To improve the quality of fruit (T.S.S.)
5. It reduces the limb breakage.

6. It promotes tree vigor and ensures more regular cropping.

7. It permits more thorough spraying and dusting of fruits during the late season application.

8. It ensures uniform ripening.

Time of thinning at blossom timing, thinning is done at marble stage. Soon after the natural fruit drop of young fruits has started.

Methods of thinning

1. Hand thinning

2. Chemical thinning

NAA at 100 ppm reduces the fruit setting from 67% to 50% in Anab-e-Shahi variety of grapes.

In mandarin, NAA 600 ppm on marble sized stage is recommended to thin the overbearing fruits so as to increase the size and quality of fruit.

9

Measurement of Soil Moisture in Fruit Orchards

Water is an important input in crop production. It acts as solvent, reactant and used for transpiration and maintaining cell turgidity. Hence, knowledge of soil moisture is essential for making irrigation scheduling. For determining soil moisture status, the commonly used methods and instruments are discussed along with their advantages and disadvantages. The soil moisture can be measured by determining either the soil water content or the soil water potential. The measurement of soil water content is an estimation of the mass or volume of water in the soil, while the soil water potential is an expression of the soil water energy status. The relation between content and potential is not universal and depends on the characteristics of the local soil, such as soil density and soil texture. We can measure soil moisture content both by direct and indirect methods. In direct method, we remove the water from the soil sample by evaporation and the amount removed is determined. This method is popularly known as gravimetric method. We measure property of soil that is affected by soil water content under the indirect methods. Indirect methods employ tensiometric, nuclear (neutron and gamma radiations), electromagnetic (radar and time domain reflectometers), hygrometric (electrical resistance, capacitance, Infrared absorption) techniques besides remote sensing for making soil water models. In the following paragraphs both the methods of soil water measurement has been discussed.

Direct Method

Gravimetric methods

In gravimetric measurement of soil moisture, we can find out the soil water content in weight and volumetric basis .It is most commonly used method of soil moisture content determination. In this method a measured weight of moist soil is oven dried at temperature between 100 and 110 °C (105 °C is typical) and the moisture removed is calculated. This temperature range has been based on water boiling temperature. Soil moisture content is expressed in percentage of weight or volume basis.

Materials

1. Augur or any other suitable tool for collecting soil samples.

2. Soil moisture can or metal container with lid.

3. Precision of ±0.01 g.

4. Hot air Oven

5. Desiccators that contains active desiccant such as magnesium perchlorate or calcium sulphate.

Procedure

1. Take the soil sample from the experimental plot at desired depths using an augar and place in soil moisture box immediately and close it with lid.

2. Weigh the soil moisture box containing moist soil sample (W).

3. Place the sample with the lid off in the oven. Adjust the temperature at 105oC, and dry for 24 hours or overnight.

4. Take the sample from the oven and place it in the desiccator until it become cool.

5. Weigh the sample along with soil moisture box and record this weight (D).

6. Return the sample to the oven and dry for several hours, cool the sample in the desiccator, and determine the weight until there is no difference between any two consecutive measurements of the weight (DS).

7. Discard the dry soil and weigh the empty soil moisture box(E)

Computations

The moisture content in dry weight basis may be calculated using the following formula:

Moist soil weight $= W-E$

Dry soil weight $= D-E$

Water content$= (W-D)-E$

Gravimetric soil moisture content ($\%\theta g$) $=$ (water content/dry soil weight) x $100g$

This may also be written as $\%\theta = \dfrac{(W - D) - E \, x100}{(D - E)}$

Water content in volumetric basis (% θv) is expressed as:

$\theta v = \theta g \; x$ bulk density of soil

Soil bulk density can be estimated by measure the volume of dry soil and the dry weight of that volume of soil and expressed as cm^3 of soil /g soil.

10

Alternate Bearing

Alternate bearing (AB) or biennial bearing (BB) is a major challenge to growers and traders of fruit crops. It is typically initiated by an abnormally heavy crop in trees (on year), followed by a light or no subsequent crop (off year). When on and off year sequence does not follow a systematic pattern it is called periodicity of cropping or irregular bearing. Alternate bearing is assigned due to genetic factor while irregular bearing may be due to lack of good orchard management practices. Alternation becomes entrained and difficult to change unless severe climatic events intervene, or drastic management interventions are made. A heavy "on" crop results in reduced vegetative shoot and root flushing, and less carbohydrate (energy reserves) build-up. Phenomenon of alternation is more prominent in the perennial fruit crops particularly in Anacardiacae (mango and pistachio), Carylaceae (Hazelnut), Oleaceae (olives), Rosaceae (apple, pear, plums, apricot etc), Rutaceae (orange, Tangor, Satsuma etc) and also tamarind, *jamun* etc. fruit crops. Within a tree species some cultivars are regular while others are alternate bearer e.g. in mango Amrapali is regular while Langra is strongly alternate bearer.

Development of Alternation

An alternate bearing cycle are due to genetical and or environmental and orchard management practices resulting in either an exceptionally heavy or a very poor (or no) crop in young trees. Prior to this, the vegetative: reproductive balance favoured vegetative growth. In certain fruit crops, during the fruit development processes high amount of photosynthates are required to be transferred from leaves to the fruits. Such enhanced photosynthetic rates of leaves near fruits cannot compensate for the high fruit "carbon" (energy) demands. Therefore, less carbon reserves are left for vegetative renewal like development of roots and initiation of new growth flushes and also development of new fruiting sites/ fruit bud differentiation processes essential for the next season's fruiting. The result is no or very low fruiting or "off" crop. It is assigned due to detrimental effect of "on" crop on the subsequent crop's flowering and fruiting. The bearing behaviour of a crop can be affected by environmental conditions, cultivar and rootstock, and management.

Alternate Bearing Index (ABI)

Pearce & Dobersek Urbanc (1967) proposed an index to measure of the extent of alternate bearing is fruits in scientific research called Alternate Bearing Index (ABI). It is calculated using a formula as given below:

ABI = (yield, year 1 - yield (years 2)/yield, year 1 + yield, year 2) in kg/tree for two consecutive (on/off) years.

It ranges from 0 (no alternate bearer) to 1 (complete alternate bearer). ABI can be expressed as a percentage by multiplying by 100.

Causes of Alternate Bearing

Broadly two causes have been assigned for alternation namely, Environmental triggers and Endogenous factors.

Environmental Factors

Several environmental triggers have been found to influence alternation like climatic stress (frost, cool weather, low air humidity). Edaphic factors (salinity, drought, water table), pests and diseases etc. Frost has more influences on terminal bearing fruits. It is more damaging during spring season. Fruit setting was found to be influenced in Valencia grown in Australia due to cool weather in November months. Excessive fruit drops were observed in olives, oranges, avocado and mango due to low air humidity during early fruit development phases. Edaphic factors such as high salinity favours leaf drop and reduction in photosynthetic area. Moisture stress during flower formation increases sterile flowers in olives, while summer drought has resulted in excessive fruit drop in pome fruits. Shallow water table (about 1 m) causes low yield in mandarin and Washington Navel oranges. Severe attack of pest and diseases devastate the whole crop and bring the trees towards alternation.

Endogenous Factors

There are several endogenous causes responsible for alternation in fruit trees such as inhibition of flower initiation by growing fruits, lack of suitable pollinizers and pollinators resulting in poor fruit set, effect of seed on prevention on fruit drop and encouragement of very heavy crop load etc. Contribution of leaves to reproductive growth, competition between vegetative and reproductive sink, fruit load, C:N ratio and imbalances of hormones are other important contributors to the alternation.

Wolstenholme (2009) of South Africa suggested a hierarchy of causative factors, controllable and uncontrollable, gradually getting closer to the proximate and ultimate factor(s) as given in Fig 1. He suggested that ultimate factor may well reside in AB gene(s), yet to be identified. Inevitably, just about every possible factor will be involved, with complex feedback and feed- forward loops and interactions. He also mentioned that a better understanding of tree growth habit and phenological cycles suggests that there are some more amenable targets for management of alternate bearing.

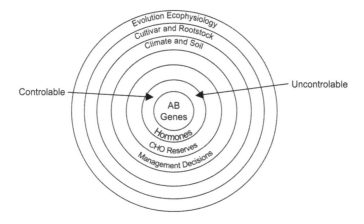

Fig. 1: Hierarchy of causative factors affecting alternation
(Source: Wolstenholme, 2009)

Control/ Management of Alternation

The basic aim of management of alternate bearing situations is to reduce overload in an „On year and induce flowering in „Off year. There are few horticultural operations which are able to reduce the intensity of alternation or irregularities in bearing.

1. Planting of Regular bearing varieties

2. **Proper orchard management:** It is necessary to follow a proper orchard management schedule by providing adequate nutrients, irrigation, proper weed, pests and disease management etc cultural operations.

3. **Pruning:** Pruning has been found to reduce alternate bearing by restoring the vegetative vigour where it is declined excessively due to over-cropping. Girdling is an ancient horticultural tool to reduce excessive vigour, and to improve flowering and fruiting - either of the whole tree or of selected branches or shoots. Girdling can be used to increase cropping in a branch renewal/rotation management system, where branches are treated as distinctive modules (mini-trees) which are re-cycled in an orderly manner by pruning.

4. **Deblossoming and Fruit thinning:** About 50 per cent of flower clusters recommended to remove soon after they emerge during „on year. This process can be done manually or by using 3-chloroisoprophyl-N-phenyl carbonate at a concentration of 250-300 ppm.

5. **Fruit thinning :** Fruit thinning is a mean to reduce the crop load in the "on" season as the overbearing favours alternation. Early hand removal of smaller fruits, before the summer drop has been shown to reduce alternate bearing phenomenon.

6. **Use of growth retardants:** Growth retardants namely, Paclobutrazole (Cultar) in mango, uniconazole in Avocado has a proven role in reducing alternation.

11

Maturity Indices, Handling and Packing Techniques

Maturity indices and their adoption at harvesting and in post harvest management of fruits provide a information against quality loss, microbial decay and deterioration of external appeal. Fruit harvesting at proper stage of maturity has direct effect on quality and market value of produce. Stage of harvesting also influence the after harvest enzymatic activities of the horticultural produce, which determines the levels of different pigments, sugars, acids and vitamins in fruits, flowers and vegetables . Maturity can be described as the attainment of a particular size, stage after which ripening takes place. However, horticultural maturity is defined as the stage of development when a plant or plant part possess the prerequisites for utilization by consumers for a particular purpose. Cucumbers, lettuce and peppers are harvested at various stages (Horticultural maturity) prior to reaching physiological maturity, or completion of growth phase.

There are many methods to determine produce maturity. The most common is the size of the individual commodity.Some other commonly used crude methods are colour change, Softening of the tissues (fig,sapota) ease of detachment from the stalk (Sapota, annona), shrivelling of fruit stalk (water melon), time elapsed from the date of flowering to picking maturity, sound by tapping jack and water melon when ripe produce hollow and dull sound on tapping but produce metallic sound if unripe etc. Another physical characteristic is the firmness and it is determined with a pressure tester. Firmness is often correlated with chemical changes that occur during ripening. Chemical characteristics include TSS, acidity, sugar, vitamins and pigments. There are certain test which accurately gives the maturity measurements such as use of colour charts (judge colour), penetrometers (firmness), brix: acid ratio etc. Other recent methods viz. thermal conductivity X-rays and NIR have also come in practice to judge the maturity of fruits and vegetables. Some commonly use maturity indices are given below in some of the important fruits.

Mango: The maturity standards have been finalized for all the commercial mango cultivars. These standards vary according to utilization purpose, mode of transportation and transit time. Baganpalli should be harvested with about

8% TSS for air transport. Most of the mango fruits are usually harvested at hard green stage when they are physiologically mature but before the onset of climacteric rise.The commonly used indices are fullness of cheeks, colour development in beak end, growth of seed hair, days after fruit set 100-120 days, specific gravity, flesh colour or flesh caratenoids, existence of powdery material called „bloom on the fruit surface, starch or dry matter content etc.

Banana: Colour, shape and firmness are the physical attributes of fruit that are frequently related to fruit maturity. Banana is harvested before full maturity in a green and hard condition. Generally in most hybrids or cultivars individual fingers are angular during early stage of development. However, as growth progresses, the finger looses angularity and become more rounded in shape in most of the banana plantains, fruits destined for distant markets are harvested at a stage known as *three quarters full* when the fingers are still clearly angular. The starch and sugar proportion of the tasteful fruit is about 1:20.

Guava: Harvest maturity of guava has been standardized based on visual appearance and destructive tests. There are a continuous decrease in specific gravity (SG), with ripe fruit reaching values <1.0.

Table : Maturity indices of some important fruits

Index	Examples
Elapsed days from full bloom to harvest	Apple, pears
Mean heat units during development	Peas, apples
Development of abscission layer	Apples, feijoas
Surface morphology and structure	Cuticle formation on grapes, Gloss ofsome fruits (development of wax)
Size	All fruits
Specific gravity	Cherries
Shape	Angularity of banana fingersFull cheeks of mangos
Textural properties	
Firmness	Apples, pears, stone fruits
Color, external	All fruits
Internal color and structure	Flesh color of some fruits
Compositional factors	
Starch content	Apples, pears
Sugar content	Apples, pears, stone fruits, grapes
Acid content, sugar/acid ratio	Pomegranates, citrus, papaya, kiwifruit
Juice content	Citrus fruits
Oil content	Avocados
Astringency (tannin content)	Persimmons, dates
Internal ethylene concentration	Apple, pears

Citrus: Firmness of different citrus fruits declines with the advancement of maturity. Colour of rind, TSS, acidity, and juice content are commonly used maturity indices for citrus. Specific gravity (SG) is not a reliable maturity index for citrus fruits. Fully mature Nagpur mandarin fruit contains about 43-44% juice, percentage of juice in kinnow varied from 43-50, Kagzi and Tahiti lime yielded 61 and 51.2 per cent juice respectively. Total soluble solids and TSS/ acid ratio are the reliable indices for assessing the maturity in citrus. Kinnow with TSS range from 13-18% considered to be the best for harvesting under Punjab conditions, Nagpur Mandarin was found best to harvest at 10-11% TSS. The ratio of TSS/acid was considered best as 22:1, 12.5:1, 16:1, 17.5:1 and 14:1 in Mosambi, Jaffa, Pineapple, Blood red and Kinnow respectively at the peak of maturity.

Litchi: The number of days from full bloom to harvest is considered to be the best maturity index for litchi fruits. All the varieties arrived at harvest maturity between 55-80 days after full bloom. The development of fruit colour is most dependable maturity index. Litchi fruits turn deep red when fully ripe and fruits harvested at this stage possess excellent quality. The maturity of fruits can also be determined by watching and touting of tubercles. The tubercles becomes slightly flattened and epicarp becomes smooth when the fruit is mature. optimum time for harvesting Bombai litchi under West Bengal condition was found between 105-110 days after anthesis i.e. second fortnight of May.

Sapota: Sapota takes around 200 days to mature from fruit set. Arrest of latex flow and change in fruit surface colour (potato colour) are the best maturity indices for sapota fruits.

Pineapple: Among the various maturity indices, specific gravity and shell colour are most practical and reliable. For fresh use, pineapple should be picked when the flesh turned into light yellow colour and flavour not too acidic. Fruits for home use should be harvested when 25% fruits showed yellowing. Under Kerala conditions, the cultivar Kew reaches at harvest maturity after 132-135 days of inflorescence emergence.

Pomegranate: Pomegranate is a non-climacteric fruit and requires maturity on the tree for its better post harvest quality attributes. The fruits are ready to harvest between 125-165 days after anthesis. At maturity, the fruit colour changes to dark yellow in summer and dark red in winter. However, the varieties like Ruby, Mridula and Bhagwa do not show much difference in colour change between fruit development and maturity period. In most of the cultivar, the buds at the anterior end of the fruit curve inwards and turned into dry black hard structures at maturity. TSS: Acid ratio and titrable acidity could be used as one of the most reliable maturity indicator for pomegranate. Varieties like Ganesh, Bassein Seedless and Alandi may be harvested at TSS, acidity and TSS: acid ratio between 8.5-10%, 0.6-0.78% and 55-60, respectively.

Papaya: The fruits of papaya showed a double sigmoid type of growth curve. Change of colour at the apical end, TSS 11-13^0 Brix, if twisted it will be easily pulled out and days after fruit set to fruit harvest are the main indices. Skin colour turning stage was attained at 130-135 days and the fruit took 155-160 days to reach eating ripe stage. The fruits harvested at colour break stage attain proper TSS (7-8%), total sugars (5-6%), acidity (0.096%) and sugar/acid rates (50-55) at ripening. For export, fruits are harvested at colour break, or between colour break and one- quarter yellow colour to obtain maximum fruit life and quality. There are others maturity indices like thermal conductivity of pulp (0.51 W/M °C for ideal ripe fruit) and NIR transmission spectra.

Aonla: In North India, 3rd week of December is the ideal harvesting time for all the commercial varieties. Fruit length and diameter increases rapidly up to 75 days after fruit set and then slowed up to 90 days when there appeared no further increase. Fruit volume increased up to 75 days after fruit set, fruit colour changed from green to yellow green or reddish green vitamin C and TSS increased up to 120 days after fruit set and this is the optimum stage for harvesting

Pear: Several indices viz., size, specific gravity, skin colour, pigments, TSS, phenolics and fruit firmness are implied to judge the maturity of pears. Fruit growth shows 3 phases-active, very active and very slow as measured by size and weight. During last 50 days near to Patharnakh pear maturity, specific gravity, starch and phenolics declined very fast while TSS, sugars and pigments showed steep spurt. Fruits reached to optimum harvest stage after 150 days of flowering when TSS, starch, phenolics and fruit firmness adjusted around 14%, 1.60%, 0.62% and 20-22.16/inch^2, respectively. Ground colour changes are from a green to yellowish green.

With ease of separation, the fruit should come off the spur easily without tearing or breaking the fruiting spur.

Peach: Peach being a highly perishable fruit, needs immediate harvesting at its maturity. In peaches the degree of maturity depends on the cultivar. Under Punjab conditions, the maturity for Flordasun and Elberta may be fixed between 71 and 74 days after fruit set when TSS (11-12%), acidity (0.48%) total sugars (11.26) and sugar/acid ratio reaches an optimum level. Fruit colour of immature fruit (light green) turned into pink with reddish tinge at maturity.

Apple: There are several methods to determine the maturity of apples. Some are often used together to best determine the harvest date of apples viz., Texture, TSS, colour, size, internal ethylene evolution, starch and heat units. Industry standards for soluble solids are at least 12 percent; for ground color a change to a yellowish cast; for iodine-starch test 60 percent of the area blue-black in color; ease of separation, the fruit should come off the spur easily without tearing or breaking the fruiting species. Generally the fruits of many commercial cultivars attend the maturity in 120-125 days after full bloom stage and it may be suggested as the suitable time for applied harvesting. Maturity indices for storage- are background color, firmness 7.5- 9.5 kg, starch score 2.5-3.5, blush more than 60% and sugar level more than 13%.

Strawberry: Harvesting is the most crucial operation in strawberry production. Skin colour, taste and TSS: acid ratio is taken in consideration to judge the maturity of fruits. Among the maturity indicators, colour of the fruits is one of the widely accepted maturity index by the growers and consumers. Berries are usually harvested when ¾ skin develops colour. For medium distance (400-500 km, 5-6 hours transportation duration) and long distance (more than 6 hours transportation duration) markets fruits are harvested at 75 and 50 per cent skin colour development stage. For local market fruits are harvested at full maturity (whole fruit colour stage). Strawberry should not be picked over ripe as colour (both internal and external) darkens during storage.

Grape: The concentration of soluble solids ($18\text{-}20^0$), pH and titrable acidity in fruits juice, easy separation of berries from the bunch and peel colour are more reliable parameters to determine the grape berry maturity and optimum harvest date. Chlorophyll, carotenoids, total phenol and polyphenol oxidase activity may also be taken as indicator for berry maturity. Among the acids, succinic, tartaric and malic acids are the main substances. Succinic acid dominates immediately after fruit set and its concentration dropped sharply as berries matured. Mallic acid content increased gradually until veraison, after which it decreased with fruit ripening.

Harvesting

The fruit growers should bestow more attention and considerable care during the picking season to reduce to a minimum level of careless handling of fruits by pickers. Different harvesting devices developed for each crops e.g. Mango harvester (IARI, IIHR, CISH, KKV etc), lime harvester, Kinnow harvester (CIPHET) etc. Picking of fruits should be done from the lower branches of a tree advancing towards the tip so that fruit dropping on ground can be avoided. During picking, care must be taken to avoid any possible damages to the branches especially to the spurs as the subsequent cropping depends upon them.Fruits should normally be harvested early in the morning. After harvesting picked fruits should be kept in shade/ cool and ventilated place to arrest respiration and break down as much as possible. Avoid any kind of damage and bruises on the fruit surface.

Handling

It includes all processes from harvesting of fruits to final disposal of produce at the consumer point. It includes the treatments given for getting the fruits ready for the market viz., packaging and wrapping, ripening and storage. One of the important treatments is the dipping the fruits in antiseptic solutions like 1-2% caustic soda to remove the dust and infestation of scale insects and washing with 1 - 1.5% of Hydrochloric acid to remove any spray residue and to improve the appearance.

Pre-Cooling

It refers to the rapid removal of the field heat from the freshly harvested fruits to slow ripening processes and reduce deterioration prior to storage and shipments. Fruit is precooled when its temperature is reduced from 3 to 6°C and is cool enough for safe transport. Precooling may be done with cold air, cold water (hydrocooling), direct contact with ice, or by evaporation of water from the product under a partial vacuum (vacuum cooling). A combination of cooled air and water in the form of a mist called hyraircooling is alsobeing used now a days.

Granding

Fruit are sorted by quality into different (two or more) grades according to country standards. It comprise of sorting product in grades or categories of quality. Two main systems exist: static and dynamic. Static systems are common in tender and/or high value crops, where produce is placed on an inspection table and sorters remove units not meeting the requirements for the grade or quality category. The dynamic system is probably much more common. In this system product moves along a belt in front of the sorters who remove units with defects. Main flow is the highest quality grade. Grading is probably one of the best practices to better price of the produce in market.

Wrapping

Covering the fruits after harvest with any material in order to improve its post harvest life is known as wrapping. The materials commonly employed as wrappers are tissue papers, waxed paper, polyfilm, cellophane paper, aluminum foils and alkathene paper etc. Wrapping minimizes the loss of moisture in shriveling, protects against the spread of diseases from one to the other, reduces bruises and other damages during transport or in storage, and makes the fruit more attractive.

Waxing

Waxing of fruits minimize moisture loss from produce surface, improves appearance and reduces the incidence of storage diseases. Wax emulsion is prepared by melting microcrystalline paraffin or cranaube wax along with emulsifiers. Citrus and apples are important fruits.

Packaging (or) Packing

The term packaging encompasses both the direct or primary packaging around the product and the secondary and tertiary packaging, the over packaging such as over warts, cartons and crates etc. The main purpose of packaging is to ensure that the product is inside a container along with packing materials to prevent movement and to cushion the produce (plastic or moulded pulp trays, inserts, cushioning pads, etc.) and for protection (plastic films, waxed liners, etc.). It needs to satisfy three basic objectives. These are to:

- Contain product and facilitate handling and marketing by standardizing the number of units or weight inside the package.

- Protect product from injuries (impact, compression, abrasion and wounds) and adverse environmental conditions (temperature, relative humidity) during transport, storage and marketing.

- Provide information to buyers, such as variety, weight, number of units, selection or quality grade, producer's name, country, area of origin, etc. Recipes are frequently included such as nutritional value, bar codes or any other relevant information on traceability.

A well-designed package needs to be adapted to the conditions or specific treatments required to be undertaken on the product. For example, if hydrocooling or ice-cooling need to be undertaken, it needs to be able to tolerate wetting without losing strength; if product has a high respiratory rate, the packaging should have sufficiently large openings to allow good gas exchange; if produce dehydrates easily, the packaging should provide a good barrier against water loss, etc. Semi-permeable materials make it possible for special atmospheres inside packages to be generated. This assists in maintaining produce freshness.

12

Fruit Drop: Types, Causes and Control Measures

Losses resulting from pre harvest and harvest drop of fruits have long been a serious problem to the fruit growers. As the fruits of some species and varieties approach the picking maturity, they tend to loosen from the stalk and considerable quantities may drop prior to and during the picking operation. Such fruits are badly damaged and if salvaged have a very low value compared with those picked from the tree. In certain fruit crops, the problem of fruit drop is very serious. Whereas in some, the fruit drop starts right from the time of fruit set and is severe at a number of stages of fruit development. In some fruit crop serious drop occurs before harvest.

Kinds of Drop

Fruitlet abscission is a common phenomenon that occurs in many plants in response to developmental and environmental cues leading to significant crop losses. There are usually three periods of fruit abscission, the first is the period of fruit set, which usually lasts for a month following full bloom also called as cleaning drop. The first drop occurs shortly after flower opening. Usually flowers with aborted pistils drop off at this stage. The second period of intense fruit drop may occur at the onset of hot summer and is referred to as "Second drop". Unfertilised flowers and some fertilized ones drop off at this stage. Some fertilized flowers also drop off as a result of adjustment in the tree between the nutritional factors and fruit set. The third drop is commonly referred to as "Post set drop". This occurs when the fruits are of „marble size due to formation of abscission layers in the young fruit stalks. This drop occurs in most deciduous fruits and this naturally thinning of fruits helps the tree to produce fruits of good size. The period of intense fruit drop may occur with the onset of hot summer and is referred as June drop or Pre harvest drop. At this stage, half developed and three-fourth developed fruits are shed due to many causes. This is the loss to the fruit grower and is a serious problem confronting the fruit growers.

Causes of Drop

The chief causes of shedding of blossom and young fruits are:

(i) Structural defects

Defects in flower parts, winter injury, spray, damage due to insect, defective pistils are responsible for early shedding of blossom in apple. Defective ovules are also responsible.

(ii) Non pollination

Self pollination may fail, while cross pollination may be prevented by lack of suitable pollen or by the absence of carriers. Even if pollen reaches the stigma, it may be washed off by rain before it can exert a stimulus.

(iii) Non fertilization

(a) Gametic sterlity

This condition seems to be common in polyploids containing an uneven multiple of the basic number of chromosomes especially in triploid varieties of apple. Such pollen grain either fails to germinate or if pollen tubes are formed these usually burst easily or they reach the ovary and give rise to unbalanced embryo.

(b) Incompatibility

Incompatibility in apple and pears is due to physiological reactions occurring between the pollen tube and the style and ovarian tissues.

(c) Failure of double fertilization

This may prevent the formation of the endosperm and thus that of the embryo.

(iv) Abortion of embryo

It may arise from genetical and unfavourable nutritional conditions and usually results in shedding of fruits.

Contributory Factors Affecting Fruit Drop

(i) Weather Condition

Heat waves followed by cold nights frequently cause excessive drop. Cool weather on the other hand tends to reduce the intensity, though it may prolong the duration of the drop or it may postpone it till its maturity. Winds blowing with rapid speed by shaking the fruit trees may loosen the fruits from their stalk and cause waves of heavy shedding.

(ii) Climatic factors

The formation of abscission layer is enhanced under high temperature, low humidity and very low temperature. In conditions of high temperature, the rate of transpiration is usually high from the leaves and fruits, consequently such fruits are unable to withstand water stress and shed easily. In mango normally the shedding of fruit is very severe when the temperature is high and humidity is low. The heaviest drop of citrus fruits known as June drop occurs when the atmospheric humidity is low and the transpiration rate is very high. Abrupt rise in relative humidity and temperature have direct relation with fruit cracking in litchi, pomegranate, cherries, lemons etc., which more often drop off from the plants in later stages.

(iii) Physiological factors

Physiological factors such as abnormal function in moisture content of the soils in the day or other causes resulting in disturbance of moisture relationship causes heavy fruit drop particularly in arid and semi arid fruit crops like ber and pomegranate.

(iv) Nutritional factors

Proper and balanced application of fertilizer is a pre-requisite for a tree to be able to carry its normal crop to maturity. Nitrogen seem to favour the early stages in setting but at later stage tends to favour shedding. Carbohydrate is important, particularly in connection with the drop of the young fruits. June drop can be prevented by a favourable supply of carbohydrate in the plant.

(v) Insect pest and diseases

Pest and disease incidence on fruit plants can result in severe shedding of blossoms and fruits. For example, mango flowers and fruits in early stages are attacked by a number of pests such as hoppers, mealy bug and diseases such as anthracnose and powdery mildew which causes heavy loss to the crop.

Fruit drop and its Control in Some Important Fruit Crops

Mango

Mango fruits drop primarily at two periods. The first drop consists of flowers and young fruitlets from anthesis to 21 days. The second drop of its young developing fruits from 28-35 days after pollination and fertilization. In third drop, fruits drop irregularly. During the first and second periods of fruit drop, high level of inhibitors and low levels of promoters appear to be major factor consisting fruit drop which have been controlled by exogeneous applications of auxins, gibberellins, cytokinins, growth retardants (cycocel, Alar) silver nitrate etc.

To control fruit drop in mango NAA or Planofix 20 ppm (0.002%) should be sprayed when the fruits are pea size.

Citrus

In citrus huge quantities of fruits have been reported to drop wherever sweet orangesgrow in Punjab. The first drop in citrus occurring during the month of May-June is mainly due to inadequate and imbalanced manuring of the trees and insufficient care.

The second drop occurring in the months of August and September is due to the attack of fungus *Collectotrichum gloeosporiodes*. As this organism is a weak parasite and becomes aggressive only on the neglected plants, under nourishment of the trees appears to be the cause.

The third drop, which is the most serious, occurs in December and January when the fruits are mature. This is caused by the fungus *Alernaria citri*. The attacked fruits ripen earlier and develop an unusual deep organge colour. The diseased fruits show a characteristic blackening of the central portion.

The use of 2, 4-D in concentration of 10 ppm and 20 ppm and in combination with Zinc sulphate (0.05%) considerably reduces the dropping of fruits. Even when used in a low concentration of 5 ppm, the 2, 4-D was found to be quite effective.

Apple

Chemicals like 2, 4-D and NAA in concentrations of 10 and 20 ppm have been found tobe effective. However, the varieties of apple have shown more specificity in this respect.

Apricot

Shipley variety of Apricot is susceptible to fruit drop. 2, 4, 5-T at 20 and 40 ppm was found effective in controlling fruit drop in apricot. Sprays of plant growth regulators are therefore recommended around the last week of march or first week of April.

Management of Fruit Drop

Planting windbreaks and shelterbelts

High wind velocity causes mechanical damage to the fruits and branches and their desiccation due to excessive transpirational loss of water. Windbreaks of fast growing and deep- rooted trees are planted around the orchard to provide a protective barrier against hot and cold winds and reduce the extent of fruit drop.

Water and Moisture Management

Moisture stress during fruit set and fruit development causes severe fruit drop in most of the fruit crops. In irrigated system, water application should be based on optimally sequenced evapo-transpiration deficits. In the deep-rooted perennial fruit trees, the concept of irrigation scheduling based on plant water content should be ideal.

Conservation of soil moisture by mulching enhances water use efficiency by reducing evaporation from soil surface as a result of cutting supply of heat energy to the evaporating site and lowering its thermal conductivity. Such practice helps in minimizing fruit drop. Mulch materials such as wheat and rice straw, sugarcane trash, sawdust, black or white polyethylene can be used for mulching.

Nutrition Management

Fertility management for controlling fruit drop should be done in such a way that the fruit trees derive most of their nutritional supplements from the residues of the intercrops. However, annual application of manures and fertilizers near the zones of maximum root activity coinciding with rainfall incidence is useful. Foliar feeding with recommended doses micro nutrients helps in over coming nutrient deficiencies.

Pest and Disease Management

The perennial nature of fruit trees provides a comparatively stable agro-ecosystem for several kinds of pests and disease to thrive and multiply. A close watch needs to be kept on build up of various pests and diseases and suitable control strategies should be initiated so as to obviate their harmful effects.

Pollinizers

The cultivar which is used as a source of pollen grain is termed as a pollinizer. Fruit crops which require pollinizers for effective pollination should be planted in adequate ratio. For example Royal delicious is used as a polliniser for Golden Delicious. In Kagzi Kalan lemon, 10% plant of pumello serve the purpose of polliniser. Likewise in aonla 10% plant of cultivar Chakiya should be planted with cultivars NA-7 and Banarasi.

Use of Drowth Regulators

Various growth regulators are recommended in different fruit crops for controlling fruit drop. However, care should be taken during preparation of solution. Recommended concentration for controlling fruit drop should be strictly followed, for example 2, 4-D at lower concentration acts as a hormone but at higher concentration may act as a weedicide.

Table: Prevention of fruit drop in different crops at a glance

Fruit crop	PGR	Concentration	Time of Application
Mango	NAA or 2, 4-D	20-30 ppm	Last week of April orwhen the fruits attain marble size.
Citrus species	2, 4-D or Gibberelic acid GA3	8-10 ppm 50 ppm	Before the young fruitsattain growth, two sprays of the recommended dose may be given.
Litchi	NAA or 2, 4-D	10-15ppm +1% ZnSo4.	At the time of fruitdevelopment
Cashewnut	2, 4-D or NAA	10 ppm	At the time of fruitdevelopment
Grape	Gibberelic acid GA3 or PCPA	100 ppm	Before harvesting
Apple	NAA or 2, 4-D or 2,4,5-T	10 ppm 20-50 ppm	Immediately after thepetals drop.

13

Determination of Specific Gravity TSS, Acidity, Ascorbic Acid and Sugar

Determination of Specific Gravity

The palatability of the citrus fruit is associated with its internal quality. Measuring specific gravity provide clues to internal quality of the fruit. The ratio of the weight of the fruit to its submerged weight indicates specific gravity. Specific gravity also indicates relationship between fruit weight and volume.

Steps: To determining specific gravity

Record the weight of fruit in air on top loading balance.

1. Fruit is placed in 1 litre beaker containing 600-700 ml water on a top loading balance.

2. The total weight is noted and the weight of submerged fruit is calculated.

3. If fruit is floating, a thin stainless steel rod with a loop at one end is used to submerge it so as more accurately record the fruit weight.

4. Calculate specific gravity as follows:

Specific gravity = weight of fruit in air ÷ weight of fruit in water

Determination of TSS

TSS refers to the total amount of soluble constitutes of the juice. These are mainly sugars, with small amount of organic acids, vitamins, proteins, free amino acids, essential oils and glucosides. TSS is an excellent guide to the sugar content of fruit. Fruit sugar levels generally increase as the fruit matures. However, levels are decrease when fruits are over mature. TSS or0Brix can be calculated either a Brix scale or a refractometer. A Brix hydrometer measures specific gravity and is calibrated to read directly in units of sugar concentration degree Brix at a temperature of 20 °C. One degree Brix is the concentration of a cane sugar solution containing 1 g of sugar/100 g of solution. The Brix reading requires temperature correction because the hydrometer is calibrated to read true only at 20 °C. A hydrometer cannot be used if juice temperature is higher the 37.5 °C.

Refractometers are quick and easy to use, especially when there are lots of samples to be tested. There are different types of refractometers. Most digital refractometrs have automatic temperature compensation for a specified temperature range (10-30 °C). Distilled or deionised water is used to set base line reading zero. This should be done regularly throughout testing. Additionally, testing solution of known °Brix level can also be made (10-30 °Brix) to more accurate calibration the instruments.

Steps

1. The fruit sample should be a minimum of 10 fruits.

2. Cut each fruit in half (at right angle of fruit axis).

3. Extract fruit juice and strain through fine sieve.

4. Put aside some juice (10 or 5 ml) for the titration.

5. Place a few drop of juice onto the stage of refractometer and take 0Brix reading.

6. If the refractometer is not temperature compensated then correct the reading and record final figure.

7. Between samples clean the refractometer with distilled water and dry.

TSS correction as per temperature

Temperature	Correction factor0Brix	Temperature	Correction factor 0Brix
10	-0.45	25.5	0.30
11	-0.40	26	0.35
12	-0.40	26.5	0.35
13	-0.30	27	0.40
14	-0.30	27.5	0.40
15	-0.30	28	0.45
15.5	-0.25	28.5	0.50
16	-0.25	29	0.55
16.5	-0.20	29.5	0.55
17	-0.15	30	0.60
17.5	-0.10	30.5	0.65
18	-0.10	31	0.65
18.5	-0.05	31.5	0.70
19	-0.05	32	0.75
19.5	0	32.5	0.75
20	0	33	0.80
20.5	+0.05	33.5	0.85
21	0.05	34	0.90

Contd.

21.5	0.10	34.5	0.90
22	0.10	35	0.95
22.5	0.015	35.5	1.00
23	0.015	36	1.0
23.5	0.20	36.5	1.10
24	0.20	37	0.15
24.5	0.25		
25	0.25		

Determining Acidity

Reagent

N/10 NaOH, Phenolphthalein indicator

Equipment

Conical flask, Burette, Pipette, Beaker

Procedure

1. Take 10 ml or 5 ml of the juice into a conical flask.

2. Add 5 drops of phenolphthalein solution (indicator).

3. Fill a burette with 50 ml of N/10 NaOH.

4. Slowly add the NaOH solution drop by drop to the flask and swirl, until the colour goes a persistent pink for at least 30 seconds (~pH 8.2).

5. Record the amount of sodium hydroxide solution used in ml.

6. Calculate acidity as under

When using 10 ml of juice: Acidity (%) = 0.064 x ml of NaOH used. When using 10 ml of juice: Acidity (%) = 0.128 x ml of NaOH used.

To convert % citric acid by volume to % citric acid by weight, divide the result by the specific gravity of the juice.

TSS / Acidity Ratio

The TSS / acid ratio is obtained by dividing the total soluble solids (°Brix corrected for acids and temperature) by the total titratable acid (% Acid, w/w) at 20°C. The 0Brix value of the fruit concerned must also be obtained before calculating sugar acid ratio. The calculation for determining the sugar / acid ratio of all produce are the same, but as some fruit contain different acid the appropriate multiplication factor must be applied to each calculation. A list of these acid and multiplication factor is as under:

Acid	Factor
Citric	0.0064 (citrus fruit)
Malic	0.0067 (apples)
Tartaric	0.0075 (grapes)

TSS acid ratio = TSS (%) ÷ Acidity (%)

TSS acid ratio of different citrus cultivars at maturity

Cultivar	Place	TSS/acid ratio	TSS (%)	Acidity (%)
Mandarin	North India	14-18	20	0.7
Nagpur mandarin	Central India	14 (spring)	17.5	0.8
		14 (Mansoon)		
Nagpur mandarin	Allahabad	10.4	8.3	0.8
Coorg mandarin	S. India	13.2 (main crop)	16.5	0.8
Kinnow	Himachal	13-14	11.0	1.2
	PradeshHaryana	10-11	9.0	1.1
	Punjab	13.7	11	0.8
Mandarin/	California	6.5 or high		
Tangerine				
Sweet orange	Florida	10.5	8.0	1.31
Hamlin,		12.0	9.8	1.23
Pineapple, Valencia				
Sweet orange	California	8.0	9.8	0.82
Mosambi	India	15	12	1.25
Pineapple	Punjab (India)	14	10.6	1.32
Jaffa	Punjab (India)	14	10.6	1.32
Blood Red	Punjab (India)	14	9.6	1.45
Valencia	Punjab (India)	10		

Determination of Ascorbic Acid

Reagents

Methaphosphoric acid pellets HPO3, Sodium 2, 6-dichlorobenzenoneindophenol, Ascorbic acid (l-Ascorbic Acid), Acetic acid, glacial

Equipment

Analytical balance, 500 ml glass stopper bottle (clear), 500 ml glass stopper bottle (amber), beaker (500 ml, 250 ml, 150 ml, 50 ml) volumetric flask (500 ml, 100 ml), burette 50 ml, pipettes (10 ml, 5 ml) Filter funnel, filter paper.

Reagent Solutions

1. Acid solution - Dissolve 10 gms. HPO3 pellets in warm distilled water (60EC) using successive portions until the acid completely dissolve, transferring portions to a 500 ml volumetric flask. Allow to cool to room

temperature and add 40 ml glacial acetic acid and sufficient distilled water to make 500 ml. (HPO3 slowly changes to H3PO4, but this solution remains satisfactorily for 7-10 days when storied in refrigerator).

2. Dye solution - 250 mg. of sodium 2,6-dicholrobenenoneindcphenol dissolve in approximately 500 ml warm (60EC) recently boiled distilled water. Cool to room temperature and transfer to brown bottle fitted with a tight stopper to protect from air and light. Store in refrigerator and filter before using. When dye solution fails to give a distinct endpoint with ascorbic acid plus acid solution, discard it.

3. Standard Ascorbic Acid Solution - Weigh accurately a 50 ml beaker and without removing from the balance pan add exactly 100 mg pure ascorbic acid. Add 10 ml acid mixture to beaker and dissolve the ascorbic acid. Transfer the solution to a 100 ml volumetric flask qualitatively, by rinsing the flask with successive portions of distilled water. Make up to volume. The solution is now exactly 1 mg of ascorbic acid per ml of solution. Make up fresh for each standardization.

4. Standardization of Dye Solution: Bring dye solution to room temperature. Transfer 10 ml aliquot of ascorbic acid solution to a 150 ml beaker and add 10 ml of the acid solution. Titrate rapidly the ascorbic acid solution with dye solution from 50 ml burette until light but distinct rose-pink persists at least 5 seconds.

Dye factor = 10 (mg ascorbic acid sued) = mg ml Dye slotion used ml

Rapid Method of Ascorbic Acid in Single Strength Orange Juice

Take a 10 ml sample of orange juice and transfer to 125 ml Erlenmeyer flask. Add 10 ml acid solution.

Titrate immediately with Standard Dye Solution to first permanent pink endpoint. First endpoint may fade due to absorbed juice in pulp.

Calculation

Ascorbic acid (mg 100 ml juice) = Dye factor x titration x 10

Rapid Method of Ascorbic Acid Determination in Concentrated Orange Juice

1. Take a 2 gm sample of concentrate and transfer to 125 ml Erlenmeyer flask. Add 10 ml acid solution.

2. Titrate immediately with Standard Dye Solution to first permanent pink endpoint.

3. Calculation: Ascorbic acid (mg 100 g concentrate) = Dye factor x titration x 100 ÷ weight of sample

A fast and reliable method for standardization of Dye Solution is as follows:

1. Use 98 to 99% pure ascorbic acid

2. Dissolve 100 mg of ascorbic acid in 100 ml of distilled water. (Use a 200 ml vol. flask)

3. Fill burette with the dye solution that is to be standardized.

4. Titrate into sample containing a little acid mixture solution and 10 ml sample of the ascorbic acid solution (use 10 ml vol. pipette for ascorbic acid solution).

5. Results of Standardization:

Determination of Sugars

Carbohydrates are polyhydroxy aldehydes or ketones or substances that yield such compounds on hydrolysis. Ratio of carbon to hydrogen to oxygen atoms is 1:2:1. The three major classes of carbohydrates are:

Monosaccharides:

Oligosaccharides

Trisaccharides

Polysaccharides

Principle

Sugars are estimayed by titrating a known amount of sugar solution against Felling s solution. The cupric ion in Felling s solution is reduced to cuprous state and precipitates as the red cuprous oxide. Only reducing sugars reduced the copper solution. Tannis, ascorbic acid and proteins act as reducing agents.

Reagent

1. Felling A: 69.28 g of $CuSO_4$ $5H_2O$ in water diluted to 1000 ml.

2. Felling B: 346 g of Rochelle salt sodium potassium tartrate and 200 g NaOH in water diluted to 100 ml.

3. Methylene blue indicator: 1% acquaous solution.

4. Potassium oxalate: 22% for neutralizing the lead acetate solution.

5. Lead acetate: 45% as clarifying agent

Neutralisation of Lead Acetate by Potassium Oxalate Solution

To determine the exact amount of potassium oxalate necessary to precipitate the lead from acetate solution, pipette 2 ml of aliquot of lead acitate solution into each of six 50 ml beaker containing 25 ml water. To the beaker add 1.6, 1.7, 1.8, 1.9, 2.0 and 2.1 potassium oxalate solution respectively. Filter each through 41 whatman filter paper and collect the filtrate in a 50 ml conical flask. To each of filtrate add a few drops of potassium oxalate solution. The correct amount of potassium required is the smallest amount which when added to 2 ml of lead acetate solution gives a negative test for lead precipitation in the filtrate. The equivalent volume of potassium oxalate should be marked on the bottle and employed when the solution is used in sugar estimation.

Standard invert Sugar Solution

Take 9.5 g of sucrose (extra pure) into a 250 ml conical flask and little quantity of water to dissolve it. Keep this mixed Fehliing solution on the burner after the addition of 40 ml of distilled water and let the content boil. When the contents start boiling, bring the burette containing hydrolysed sugar solution and titrate slowly in a drop wise manner till the solution turns to brick red colour. Discontinue the titration at this point and add 2-3 drops of methylene blue indicator and titrate till the blue colour changed to brick red colour. Note the titre value and make the calculation.

Procedure

A. Standardization of Fehlling's solution

Take 5 ml each of Fehliing A and B solution with separate pipettes in a 25 ml conical flask. Keep this mixed Fehlling solution on the burner after the addition 40 ml of distilled water and let the content boil. When the content starts boiling, bring the burette containing hydrolysed sugar solution and titrate slowly in a drop wise manner till the solution turns to brick red colour. Discontinue the titration at this point and add 2-3 drops of methylene blue indicator and titrate till the blue colour changed to brick red colour. Note the titre value and make the calculation.

B. Derivation of the factor

9.5 g of sucrose = 10 g glucose and fructose

i.e 10 g of glucose and fructose/1000 ml or 1g 100 ml

100 ml of 1% sucrose solution contains 1 g sugar

25 ml of 1% sucrose solution contains = 25/100 = 0.25 g

This 25 was again made to 100 ml

100 ml will thus contain = 0.25 g sugar

22.45 ml of neutrilised sucrose contains 22.45/100x 0.25 = 0.056 of sugar

22.45 ml of neutrilised sucrose has reduced 10 ml of Fehlling solution (A and B).

10 ml of Fehlling solution = 0.056 g of sugar

Hence the factor of Fehlling solution is 0.056

Estimation of sugar

Take 10 g of sample and grind in mortar and pestle with inert sand, dilute to suitable volume and clarified by addition of 45% neutral lead acetate solution. For juices 10 ml of sample can be taken directly.

Neutrilise the excess lead acetate solution by addition of 22% of potassium oxalate solution. Make up the volume with water and filter to get a clarified solution.

In the clarified filterate determine the reducing sugar by titrating against Fehlling solutions using methylene as an indicator as described above.

Calculation

(i)   Reducing sugar (%) = $\dfrac{\text{mg of invert sugar x dilution x 100}}{\text{Litre volume x wt or volume of sample x 1000}}$

(ii)   Total sugar (%) = calculate as in 90 making use of the titre value obtained in the determination of total sugars after inversion.

   Note: the total sugars in the sample are estimated by hydrolising the clarified solution with concentrated HCl acid. The hydrolyzed solution is then neutrilised and titrated as in the case of reducing sugars.

(iii)   % non reducing sugars (% sucrose) = (% total invert sugars - % reducing sugars) x 0.95

(iv)   % total sugars = % reducing sugars - non reducing sugars

Estimation of Reducing Sugar by Dinitrosalicylic Acid Method

For sugar estmation an alternative to Nelson-Somogyi method is the dinitrosalicylic acid method – simple, sensitive and adoptable during handling of a large number of samples at a time.

Materials

Dinitrosalicylic Acid Reagent (DNS Reagent):

Dissolve by stirring 1g dinitrosalicylic acid, 200mg crystalline phenol and 50mg sodium sulphite in 100mL 1% NaOH. Store at 4°C. Since the reagent deteriorates due to sodium sulphite, if long storage is required, sodium sulphite may be added at the time of use.

40% Rochelle salt solution (Potassium sodium tartrate).

Procedure

1. Weigh 100 mg of the sample and extract the sugars with the hot 80% ethanol twice (5mL each time)

2. Collect the supernatant and evaporate it by keeping it on a water bath at 80°C.

3. Add 10mL water and dissolve the sugars.

4. Pipette out 0.5 to 3mL of the extract in test tubes and equalize the volume to 3mL with water in all the tubes.

5. Add 3mL of DNS reagent.

6. Heat the contents in a boiling water bath for 5min.

7. When the contents of the tubes are still warm, add 1mL of 40% Rochelle salt solution.

8. Cool and read the intensity of dark red color at 510nm.

9. Run a series of standards using glucose (0 to 500mg) and plot a graph.

Calculation

Calculate the amount of reducing sugars present in the sample using the standard graph.

References

Miller, G L (1972) Anal Chem 31 426.

Ranganna, S. (1986). Handbook of Analysis and Quality control for Fruit and Vegetable Products. Tata Mcgraw Hill Publishing Co., New Delhi.

Lehninger, A.L. (1976). Bichemistry, CBS publishers and Distributors, New Delhi.

Lane, J.H. and Eynon, J. (1923). Determination of reducing sugars by Fehliing s solution with methylene blue as an indicator. *J. Sco. Chem. Ind.* 42: 32.

14

Nutrient Assessment in Fruit Crops

Management system adopted for fruit trees are quite different, especially with regard to nutrient than that adopted for field crops. The shift in management system could be commenced 3-4 years after planting depending on the tree size. Nutrients are required for flowering and fruiting while at the same time trees are allowed to grow and maintain sufficient vigour for producing high yields in following years. To maintain productivity of trees in long run and maintain the sustainability of tree production capacity, application of nutrient should be based on actual requirement and availability of nutrient in the soil. Application of plant nutrients economically at correct time with right amounts in a way that nutrients could be taken up by plants efficiently with minimum losses. The purpose of assessment of nutrient requirement of fruit trees is to keep mineral nutrient levels in the tree with in the desired range to have the growth and development effects and fruiting of trees as desired.

The most important thing in nutrient management for bearing trees is to analyze the importance of timing of nutrient application in relation to tree phenology or growth cycle.

Factors influencing nutrient content of the soil

Type of vegetation cover; e.g. with legume cover there could be higher N, etc.,

Application of manures and fertilizers. Application of soil amendments. Inheriting status of soil

Factor affecting nutrients content of the leaf

Fertilizers input

Type of vegetation cover; e.g. with legume cover there could be higher N, etc.,

Factors inherent in the tree; genetic make-up, age of crop, time of sampling; yearly (seasonal) variation, age of leaves, position of leaves, yield of crop.

Leaf Samples

Leaf analysis is an important tool to estimate nutritional requirement of fruit trees. Leaf nutrient content can be obtained by analysing leaf tissues at proper growth stage. Thereafter, nutrient requirement of fruit trees can be calculated based on optimum norms of tissue nutrient if different fruit crops.

Plant tissues sampling guideline for fruit crops

Sl.No.	Crop	Index tissues	Growth stage
1.	Banana	Petiole of 3rd open leaf from apex	Bud differentiation stage
2.	Custard apple	5th leaf from apex	2 months after new growth
3.	Fig	Fully expanded leaves, mid shoot current growth	July-August
4.	Grape	5th petiole from base	Bud differentiation stage for yield fore cast. petiole
5.	Citrus	3 to 5 month old leaf from new flush. 1st leaf of the shoot	June
6.	Guava	3rd pair of recently matured leaves	Bloom stage
7.	Mango	Leaves + petiole	4 to 7 months old leaves from middle of the shoot
8.	Papaya	6th petiole from the apex	6 month after planting
9.	Pomegranate	8th leaf from pex	Bud differentiation. In april for February crop and August for June crop
10.	Sapota	10th leaf from apex	September
11.	Phalsa	4th leaf from apex	One month after pruning
12.	Ber	16th leaf from apex from secondary or tertiary shoot	Two month after pruning

Calculation of quantity of fertilizers based on nutrient requirement

Quantity of fertilizer = quantity nutrient required x 100 ÷ nutrient present in fertilizer

Or

$Q = N1 \times 100 \div N2$

Where

Q = Quantity of fertilizer

N1 = quantity nutrient required

N2 = nutrient present in fertilizer Or

Q = quantity of nutrient required x factor

Calculation of Quantity of Complex Fertilizers

First find out the quantity of fertilizer required for supplying of whole quantity of major nutrient present in the particular fertilizer. Thereafter, calculate the amount of second nutrient supplied through the calculated quantity of fertilizer. Then subtract calculated amount of nutrient from the whole amount of second nutrient. Calculate the quantity of another source of fertilizers for balance nutrient quantity of second nutrient.

Calculation of quantity of chemicals for spray solution

$$V_1 = \frac{C2 \times V2}{C1}$$

Where;

V1 = quantity of chemical or commercial product

V2 = volume of spray solution to be prepared

C1 = nutrient content in chemical or commercial product

C2 = concentration of spray solution

Or

$$\text{Quantity of chemical t} = \frac{\text{Qty of spray solution required} \times \text{conc. of solution desired}}{\text{\% of actual ingredient of nutrient in chemical}}$$

Nutrient content in different fertilizers

Fertilizer	N (%)	P_2O_5 (%)	K_2O (%)	Ca	Mg	S
Urea	46.0					
Calcium Ammonium Nitrate	26.0					
Ammonium Nitrate	34.0					
Ammonium chloride	25.0					
Calcium Nitrate	15.5					
Ammonium sulphate	20.6					24.0
Single super phosphate		16.0				
Rock phosphate		18.0				
Bone meal	2	20		25		
Potassium sulphate			52			16
Potassium chloride			60			
Potassium Magnesium Sulphate			22			

Liquid fertilizers

Liquid Fertilizers	Nitrogen (%) by wt.	Water soluble P_2O_5 (%) by wt.	Water soluble K_2O (%) by wt.	Ca	Mg	S
Urea Ammonium Nitrate	32					
Superphosphoric Acid		70				
Ammonium Polyphosphate	10	34				
Mix fertilizers						
DAP	18	46				
Potassium Nitrate	13		45			
NPK (15:15:15)	15	15	15			
NPK (10:26:26)	10	26	26			
NPK (12:32:16)	12	32	16			
NPK (20:10:10)	20	10	10			
Manure						
Cattle	1.5	1.5	1.2	1.1	0.3	
Poultry						
Broiler litter	3.0	3.0	2.0	1.8	0.4	0.3
Hen-litter	1.8	2.8	1.4			
Sheep manure	0.6	0.3	0.2			
Sewage sludge	5.0	6.0	0.5	3.0	1.0	1.0

Micronutrient content in different chemicals

Commercial product	Content (%)	Commercial product	Content (%)
Cu EDTA	13 Cu	Ferrous sulphate	20% Fe
Copper sulphate	35 cu	Fe EDTA	12% Fe
Borax	11% B	Manganese sulphate	24% Mn
Boric acid	17% B	Ammonium molybdate	54% Mo
Calcium borate	10% B	Sodium molybdate	46% Mo
Sodium tetraborate	14% B	Zinc chelate	14% Zn
Magnesium borate	21% B	Zinc sulphate	36% Zn
Basic slag	13% Fe; 3% Mn		

Factor for calculating quantity of different fertilizers

Fertilizers	Factor	Fertilizers	Factor
Nitrogenous fertilizers		Phosphatic fertilizers	
Urea	2.17	Single super phosphate	6.25
Calcium Ammonium Nitrate	3.85	Rock phosphate	5.56
Ammonium Nitrate	2.94	Bone meal	5.00
Ammonium chloride	4.0	**Potassic fertilizers**	
Calcium Nitrate	6.45	Potassium sulphate	1.92
Ammonium sulphate	4.84	Potassium chloride	1.66
		Potassium Magnesium Sulphate	4.55

Optimum norms for horticultural crops

Nutrient	Grape (Thompson SL)	Mango (Totapuri)	Acid lime	Pomegranate	Banana (Robusta)
N (%)	0.87-1.61	0.84-1.53	1.53-2.10	0.91-1.66	1.67-343
P (%)	0.29-0.65	0.64-0.15	0.10-0.15	0.12-0.18	0.12-0.21
K (%)	2.0-3.0	0.52-1.10	0.96-1.66	0.61-1.59	2.28-4.14
Ca (%)	0.98-1.36	1.97-3.20	3.05-3.43	0.77-2.00	0.48-1.70
Mg (%)	0.63-1.10	0.40-0.65	0.40-0.60	0.16-0.42	0.33-0.58
S (%)	0.09-.13	0.15-0.22	0.25-0.29	0.16-0.26	0.03-0.18
Fe (ppm)	54-80	48-86	117-194	71-214	53-196
Mn (ppm)	42-209	57-174	21-63	29-89	112-417
Zn (ppm)	30-88	25-33	25-50	14-72	8-38
Cu (ppm)	5-10	3.1-8.0	8.7-14.8	29-72	10-32

15

Major Diseases of Fruit Crops and Their Remedial Measures

Diseases are often the most important constraint to the production of different fruit crops. They directly or indirectly reduce yields by debilitating the plant, and depleting the fruit quality. They range from aesthetic problems that lower the marketability of the harvested product to lethal problems that devastate the overall production. Virtually all fruit crops are affected by one or more serious diseases. Diseases determine how and where a crop is produced, what post-harvest treatments are utilized, in what markets the crops are sold, and whether production is sustainable and profitable. Hence, as a fruit crop student one should know have the knowledge of different diseases occurring on major fruit crops, their symptoms and management strategies;

Mango

Anthracnose (*Colletotrichum gloeosporioides*)

Symptoms

- Produces brown to dark brown leaf spots, blossom blight, wither tip, twigs blight and fruit rot.

- Small blister like spots develop on the leaves and twigs. Young leaves wither and dry. Affected branches ultimately dry up.

- Black prominent spots appear on fruits, the pulp of which become hard, crack and start to decay at ripening.

- Infected fruits drop.

Management

- Spray *Psuedomonas fluorescens* at three weeks interval commencing from March @ 5 g/ l on flower branches. About 5-7 sprays one to be given on flowers and bunches.

- After harvest the fruits are treated with hot water (50-55°C) for 15 min. or dip in Benomyl solution (500 ppm) or Thiobendazole (1000 ppm) for 5 min. followed by air-drying.

Powdery mildew (*Oidium mangiferae*)

Symptoms

- It attacks the leaves, flowers, panicle stalks and fruits

- Shedding of infected leaves occurs when the disease is severe.

- The affected fruits do not grow in size, turn dark and then drop before attaining pea size.

- Survives as dormant mycelium on affected leaves and cloudy weather along with cold favours it attack.

Management

- Dusting the plants with sulphur powder (250-300 mesh) at the rate of 0.5 kg/ tree.

- The first application may be soon after flowering, second 15 days later (or) spray with wettable sulphur (0.2%), (or) karathane (0.1%).

- One prophylactic spray must before panicle emergence in north Indian plains to avoid the future attack.

Stem end rot (*Diplodia natalensis*)

Symptoms

- It appears on fruit as dark epicarp around the base of the pedicel as a circular, black patch.

- High humidity conditions lead to its rapid spread on the whole fruit within two or three days.

- The pulp becomes brown and somewhat soft and watery.

- Inoculum perpetuate on dead twigs and bark of the trees, spread rapidly by rains at full maturity.

Management

- Prune and destroy infected twigs.

- Spray Carbendazim or Thiophanate Methyl (0.1%) or Chlorathalonil (0.2%) at fortnightly interval during rainy season at two to three times.

Red-rust (*Cephaleuros virescens*)

Symptoms

- Caused by algal attack foliage and young twigs.

- Rusty spots appear on leaves, initially circular, slightly elevated, coalesce to form irregular spots.

- The spores mature fall off and leave cream to white velvet texture on the leaf surface.

Management

- Spray Bordeaux mixture (0.6%) or copper oxychloride (0.25%).

Sooty mould (*Capnodium mangiferae*)

Symptoms

- Black encrustation is formed which affect the photosynthetic activity.

- The fungi produce mycelium, which is superficial and dark. These growths take place on sugary secretions made by the plant hoppers, jassids, aphids and scale insects.

Management

- Controlling the insects by spraying systemic insecticides like monocrotophos or methyl dematon 0.2% or 2 ml/l.

Banana

Panama disease (*Fusarium oxysporum* f.sp. *cubense*)

Symptoms

- Yellowing of the lower most leaves starting from margin to midrib of the leaves.

- Yellowing extends upwards and finally heart leaf alone remains green for some time and which eventually also get affected.

- The leaves break near the base and hang down around pseudostem.

- Longitudinal splitting of pseudostem. Discolouration of vascular vessels as red or brown streaks.

- Commonly spread through use of infected rhizomes.

- Spread with flood irrigation.

Management

- Avoid growing of susceptible cultivars *viz.*, Rasthali, Monthan, Red banana and Virupakshi.

  Grow resistant cultivar Poovan, Grand Naine etc.

- Treatment of sword suckers with Carbofuran (3 g /rhizome) granules followed with dip slurry of 2% Carbendezim followed by air drying in shade for 2-3 h.

- Adopt tissue culture raised plants.

Moko disease (*Pseudomonas solanacearum*)

Symptoms

- Leaves become yellow and progress upwards. The petiole breaks and leaves hang.

- When it is cut open, there is discolouration in vascular region with pale yellow to dark brown colour. There is discolouration is in the central portion of the corm and pseudostem, when cut transversely give bacterial ooze.

- Internal rot of fruits with dark brown discoloration.

Management

- Avoid collection of plant material from sick plantations.

- Practice soil solrozation during summer.

- Fallowing and crop rotation to be followed

- Disinfection of pruning tools.

- Providing good drainage after heavy flash rains.

Tip over or Heart rot (*Erwinia carotovora* subsp. *carotovora*) Symptoms

- The base of the pseudostem and upper portion of the corm are affected and leads to rotting.

- Young 1-3 month-old plantation susceptible during summer months.

Management

- Plant disease free suckers.

- Immediate removal of infected plants and burn or burry in pit.

- Drench rhizosphere soil with methoxy ethyl mercuric chloride (Emisan-6) 0.1% or sodium hypochlorite (10% v/v) or bleaching powder (20 g/litre/tree).

Sigatoka or leaf spot (*Mycosphaerella musicola* & *Cercospora musae*)
Symptoms

- On leaves small light yellow or brownish green narrow streaks appear. They enlarge in size and become linear, oblong, brown to black spots with dark brown brand and yellow halo.

- Black specks of fungal fruitification appear on the affected leaves followed by drying and defoliation.

Management

- Removal and destruction of the infected leaves.

- Spray propiconazole 0.1%+ carbendazim 0.1% or chlorothalonil 0.25% along with wetting agent such as Teepol @1ml/l water.

Banana bunchy top (*Banana bunchy top virus*) Symptoms

- The infected plants are dwarfed, leaves have dark broken bands of green tissues on the veins, leaves and petioles, size is reduced along with marginal chlorosis and inward curling, growing upright and become brittle.

- Several leaves get crowded at the top with no shooting and bunch production.

- The disease is transmitted primarily by infected suckers and also by aphid vector *Pentalonia nigronervosa*

Management

- Select suckers from disease-free areas/ plantations.

- Control of aphid vector by spraying methyl demoton 1 ml/l or monocrotophos 2 ml/l or phosphomidon 1 ml/ l or injection of monocrotophos 1 ml/plant (1 ml diluted in 4 ml).

- Infected plants are destroyed using 4 ml of 2,4-D (50 g in 400 ml of water).

Banana streak virus

- Disease severity is very variable, and probably depends on environmental conditions, as well as on host and virus genotypes.

- The most characteristic foliar symptoms of infection are chlorotic streaks, which become necrotic with time. The leaf lamina may also be narrower, thicker and become torn.

- Stunting of the plant, constriction of the bunch on emergence (choking), altered phyllotaxis (leaves arranged in a single vertical plane instead of the normal spiral pattern), and detachment and splitting of the outer leaf sheaths of the pseudostem.

Management

- The eradication of infected plants followed by burning.

- The use of BSV-free planting material.

- BSV can be carried in *in vitro* plantlets, hence indexed and certified plants should be procured.

Banana bract mosaic virus

Symptoms

- The name is derived from the conspicuous discoloration and necrotic streaks that develop on the bracts of the male bud.

- Early symptoms take the form of greenish to brownish spindle-shaped streaks irregularly scattered along leaf petioles. As the disease progresses,

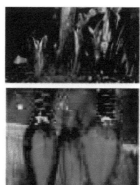

similar discolorations become very marked on the bracts of the male inflorescence, the fruit bunch, and even on the fruits themselves.

- A diagnostic symptom of the disease is the spindle- shaped streak formed on the pseudostem after removal of dried leaf sheaths.

Management

- Effective control of the disease is similar to that of other viral diseases.

- Use of virus-free propagules which are certified.

Guava

Anthracnose (*Colletotrichum gloeosporioides*)

Symptoms

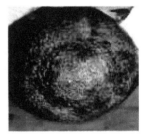

- Sunken, dark coloured, necrotic lesions appear on fruits. Under humid conditions, the necrotic lesions become covered with pinkish spore masses.

- As the disease progresses, the small sunken lesions coalesce to form large necrotic patches affecting the pulp.

- The fruit loose marketability.

Management

Spray Mancozeb (0.25%).

Papaya

Stem rot / Foot rot (*Pythium aphanidermatum*)

Symptoms

- Water soaked spots on the stem at the ground level which enlarge and girddle the trunk turns brown or black and rot, followed by terminal leaves turning yellow which later droop off. In later stages entire plant topples over and dies.

- Favoured by standing water or rain.

Management

- Seed treatment with Thiram or Captan (4 g/kg)
- Soil drenching with chlorothalonil or copper oxychloride (0.25%) or Bordeaux mixture (1%) or Metalaxyl (0.1%).

Powdery mildew (*Oidium caricae*)

Symptoms

- White mycelial growth appears on the upper leaf surface, flower stalks and fruits. Severe attack leads to yellowing and defoliation.

Management

Spray wettable sulphur (0.25%) or Dinocap (0.05%) or

Chinomethionate (0.1%) **or** Tridemorph (0.1%).

Papaya ring spot (*Papaya ring spot virus*)

Symptoms

- Vein clearing, puckering and lobbing of leaf tissues. Margin and distal parts of leaves roll downward and inwards, mosaic mottling, dark green blisters, leaf distortion, which result in shoe string formation, stunting of plants.

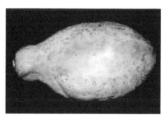

- On fruit surface circular concentric rings are produced. If affected at early stage no fruit formation takes place.
- Spread by vector *Aphis gossypii* and *A. craccivora*.

Management

- Raise papaya seedlings under insect-proof net houses.
- Raising sorghum / maize as barrier crop before planting papaya.
- Rogue out affected plants immediately.
- Avoid growing cucurbits around the field.

Leaf curl (*Papaya leaf curl virus*)

Symptoms

- Curling, crinkling and distortion of leaves, reduction of leaf lamina, rolling of leaf margins inward and downward, thickening of veins.

- Leaves become leathery, brittle and distorted. Plants are stunted. Affected plant does not produce flowers and fruits.

- Spread by vector whitefly (*Bemisia tabaci*).

Management

- Uproot and immediate burning of infected plants.

- Avoid growing tomato, tobacco near papaya plantations.

- Spraying with systemic insecticides to control the vector.

Anthracnose (*Colletotrichum gloeosporioides*)

Symptoms

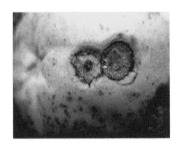

- It appears as necrotic spots on leaf and stem. Initially brown superficial discoloration appear on the skin, which are circular and slightly sunken.

- Then they coalesce in which sparse mycelial growth appear on the margin of a spot.

- Under humid condition salmon pink spores are released and the fruits get mummified and deformed.

Management

- Spray with carbendazim (0.1%) or chlorothalonil (0.2%) or Mancozeb (0.2%).

Citrus

Gummosis (*Phytophthora parasitica, P. palmivora, P. citrophthora*)
Symptoms

- Initial symptoms are dark staining of bark which progresses into wood formation.

- Bark at the base of the trunk is destroyed resulting in girdling and finally death.

- Bark in infected areas dries, shrinks, cracks and shreds lengthwise in vertical strips resulting in profuse exudation of gum.

- Infection extends to crown roots.

- Spread rapidly with standing water when applied as flood irrigation in water logged areas and heavy soils.

- Sporangia spread by splashing rain water, irrigation water and wind.

Management

- Mechanical injuries to crown roots or base of the stem during cultural operations should be avoided.

- If lesion has girdled less than half the girth, remove the diseased bark with a knife along with half an inch of the healthy bark.

- Bark of trunk should be coated with Bordeaux paste.

- Disinfection of tools used in cultural operations.

Scab/Verucosis (*Elsinoe fawcetti*)

Symptoms

- Attacks leaves, twigs and fruits of mandarin. Sour orange, lemon, mandarin, tangelos extremely susceptible; grapefruit, sweet orange and acid lime highly resistant. Severe in rainy seasons.

- On the leaves the disease starts as small pale orange coloured spots followed by distortion into firm hollow conical growth with lesion at the apex.

- The crest of this growth becomes covered with scabby corky tissue colour at first but later becomes dark olive with age.

- Lesions also appear on undersurface of leaf. They penetrate leaf and are later visible on both sides.

- Infected areas run together and cover large area. Leaves wrinkled, distorted and stunted. On twigs also similar lesions are produced.

- They form corky outgrowths. On fruits irregular scab spots or caked masses produced.

- Cream colour in young fruits; dark olive grey on old fruits.

- Fruits attacked when young become misshapen with prominent warty projections. They drop prematurely.

Management

Spray carbendazim (0.1%) at fifteen day interval for at least three times.

Canker (*Xanthomonas campestris* pv. *citri*)

Symptoms

- Acid lime, lemon and grapefruit are most affected. Rare occurance on sweet oranges and mandarins.

- Affects leaf, twig and fruits.

- Lesions are typically circular with yellow halo; appear on both sides of leaf, severe in acid lime. When lesions are produced on twigs, they are girdled and die.

- On fruits, canker lesions reduce appearance quality.

Management

- Streptomycin sulphate (500-1000 ppm); or Phytomycin (2500 ppm) or copper oxychloride (0.2%) at fortnight intervals.

- Control leaf miner when young flush is produced.

- Prune badly infected twigs before the onset of monsoon.

Tristeza or quick decline (*Citrus tristeza virus* -CTV)

Symptoms

- Lime is most susceptible both as seedling or buddling on any rootstock.

- But mandarin and sweet orange seedlings on rough lemon, trifoliate orange, citrange; Rangpur lime rootstocks tolerant; susceptible root stocks are grapefruit and sour orange.

- In sweet orange or mandarin on susceptible rootstocks, leaves develop deficiency symptoms and abscise.

- Roots decay, twigs die back. Fruit set diminishes; only skeleton remains.

- Fine pitting of inner face of bark of sour orange stock.

- Grapefruit and acid lime are susceptible irrespective of rootstock.

- Acid lime leaves show large number of vein flecks (elongated translucent area).

- Tree stunted and dies yield very much reduced. Fruits are small in size.

- Use of infected bud wood and *Toxoptera citricida* (aphid) is the important vector.

Management

- For sweet orange and mandarin, avoid susceptible rootstocks.

- For acid lime, use seedlings pre-immunized with mild strain of tristeza.

Exocortis of scaly butt (*Viroid*)

Symptoms

- Affects mostly the Rangpur lime, trifoliate orange and citrange rootstocks.

- Vertical cracking and scaling of bark on the entire rootstock along with stunting of plant.

Management

- Spray of systemic insecticide like monocrotophos (o.2%) to control the aphid vector.
- Use virus-free certified budwood and tolerant rootstocks like rough lemon
- Periodically cleaning of budding knife with 0.1% disodium phosphate solution.

Greening (*Liberobactor asiaticum*)

Symptoms

- Stunting of leaf, sparse foliation, twig die back, poor crop of predominantly greened, useless fruits.
- Foliar chlorosis, mottling resembling zinc deficiency often seen.
- Young leaves appear normal but soon assume on outright position, become leathery and develop prominent veins and dull olive green colour. Green circular dots appear on leaves with twigs becoming upright and produce smaller leaves.
- Fruits small, lopsided with curved columella. The side exposed to direct sunlight develops full orange colour but the other side remain dull olive green.
- Fruits with low juice and soluble solids, high in acid unfit for processing. Seeds poorly developed, dark coloured and aborted.
- Spread through infected budwood and psyllid vector -*Diaphorina citri.*

Management

- Control of vectors with insecticides.
- Use pathogen-free bud wood for propagation.
- 500 ppm tetracycline sulphate sprays at 15 day intervals.

Grape

Downy mildew (*Plasmopara viticola*)

Symptoms

- Irregular, yellowish, translucent spots on the upper leaf surface and powdery growth on lower surface.
- Affected leaves become yellow, brown, dry and cause premature defoliation.

- Dwarfing of tender shoots, with brown, sunken lesions on the stem.

- White growth of fungus on berries, which subsequently becomes leathery and shrivels.

- Infection of berries results in soft rot formation with no cracking.

- Optimum conditions for occurrence as 20-22°C temp and relative humidity of 80-100 per cent.

Management

Spray Bordeaux mixture (1%) or Ridomil (Metalaxyl + Mancozeb) 0.4%.

Powdery mildew (*Uncinula necator*)

Symptoms

- Formation powder like powdery like growth on upper surface of the leaves followed by malformation and discolouration.

- Discolouration of stem to dark brown.

- Floral infection cause shedding of flowers and poor fruit set.

- Early berry infection results in shedding of affected berries. At later stages appear as powdery growth and cracking of skin.

- Occurrence is almost certain during sultry warm conditions with dull cloudy weather.

Management

- Spray wettable sulphur (0.25%) or Chinomethionate (0.1%) or Dinocap (0.05%).

Bird's Eye Spot/Anthracnose (*Gloeosporium ampelophagum Elsinoe amphelina*)

Symptoms

- It appears first as dark red spots on the berry, which then become circular, sunken, ashy-gray and in later stages get surrounded by a dark margin.

- The fungus also attacks shoots, tendrils, petioles, leaf veins and the fruit stems. The spots gradually unite and girdle the stem, causing death of the tips.

- Commonly appear on warm wet weather in low lying and badly drained soils.

Management

- Clipping of infected twigs.

- Spray of copper oxychloride (0.2%) or Mancozeb (0.25%)

Fruit Spot (*Cercospora* sp.)

Symptoms

- The affected fruits show small irregular black spots, which later on coalesce, into big spots.

Management

- The diseased panicles should be collected and destroyed.

- Two to three spray at 15 days interval with Mancozeb (0.25%).

Pomegranate

Fruit Blight (*Colletotrichum gloesporioides; Pseudocercospora punicae & Cercospora punicae*)

Symptoms

- The disease is characterized by appearance of small, irregular and water-soaked spots on leaves and developing fruitlets. Affected leaves fall off.

Management

Spraying Mancozeb (0.25%) at 15 days interval gives good control of the disease.

Fruit spot (*Alternaria alternata*)

Symptoms

- Small reddish brown circular spots appear on the fruits.
- As the disease advances these spots, coalesce to form larger patches and the fruits start rotting.
- The arils get affected which become pale and become unfit for consumption.

Management

- The affected fruits should be collected and destroyed.
- Spraying Mancozeb (0.25%) effectively control the disease.

Ber

Powdery mildew (*Oidium erysiphoides* f.sp. *zizyphi*)

Symptoms

- The developing young leaves show a white powdery mass causing them to shrink and defoliate. On fruitlets small, white powdery growth appears which later enlarge and coalesce and finally turn brown to dark brown.
- Affected young fruits drop off prematurely or become corky, cracked mis-shapen and underdeveloped.
- Matured fruits turn rusty and are rendered unmarketable.

Management

Spray dinocap (0.05%) or wettable sulphur (0.25%) during first and third weeks of November or when the fruit attains pea size.

Alternaria leaf spot (*Alternaria chartarum*)

Symptoms

- Formation of small irregular brown spot on the upper surface of the leaves and dark brown to black spots on lower surface.

- The spots coalesce to form big patches. The diseased leaves later drop.

Management

- The disease can be controlled effectively by spraying mancozeb (0.25%).

Rust (*Phakopsora zizyphi-vulgaris*)

Symptoms

- On the lower surface of the leaves small, irregular, reddish-brown pustules appear which may cover the entire area

- The infected leaves dry and defoliate.

Management

- Spray Mancozeb (0.2%) or Zineb (0.2%) or wettable sulphur (0.2%).

Soft rot (*Phomopsis natsume*)

Symptoms

- Appearance of light russet vinaceous coloured, irregular spots on the fruits, which increase in size and make the whole fruit into pulpy, brown to black in colour with soft and loose outer skin.

Management

- Spray carbendazim (0.05%).

Apple

Scab (*Venturia inaequalis*) Symptoms

- On lower side of the leaf lesions appear as olivaceous spots which turn dark brown to black and become velvety.

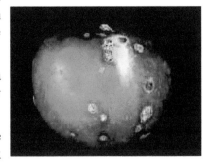

- On young foliage, the spots have a radiating appearance with a feathery edge.

- On older leaves the lesions are more definite in outline forming convex surface with corresponding concave area on the opposite side. In severe infection, the leaf blade curved, dwarfed and distorted.

- Fruits show small, rough, black circular lesions. The centre of the spots becomes corky with yellowish halo around the lesions.

Management

- Follow clean cultivation practices.

- Collection and destruction of fallen leaves.

- Spray Tridemorph (0.1%) before flowering; Mancozeb (0.25%) at bearing stage; 5% urea prior to leaf fall in autumn and 2% before bud break to hasten the decomposition of leaves. The fungicidal spray is as given below;

S.No.	Tree stage	Fungicide per 100 litre water
1.	Silver tip to given tip	Captafol 200 g (or) Captan 300 g or Mancozeb 400 g
2.	Pink bud or 15 days after 1st spray	Captan 250 g or Mancozeb 300 g
3.	Petal fall	Carbendazim 50 g
4.	10 days later	Captan 200 g or Mancozeb 300g
5.	14 days after fruit set	Captofol 150 g

Powdery mildew (*Podosphaera leucotricha*)

Symptoms

- Small patches of white powdery growth appear on upper side of leaves, which may spread to both the sides.

- Twigs are also infected. Affected leaves fall off in severe infection.

- Fruit buds are also affected and deformed or remain small.

Management

- Spray Dinocap (0.05%) or Chinomethionate (0.1%).

Fire blight (*Erwinia amylovora*)

Symptoms

- The initial symptom usually occurs on leaves, which become water soaked, then shrivel turn brownish to black in colour and fall or remain hanging on the tree.

- The symptom spread to twigs. Terminal twigs wilt from tip to downward and also spread to branches.

- Fruits become water soaked, turn brown, shrivel and finally becomes black with oozing of water.

Management

- Removal and destruction of affected plant parts.

- Removal of blighted twigs

- Spray 500 ppm streptomycin sulphate.

Soft rot (*Penicillium expansum*)

Symptoms

- Young spots start from stem end of fruit as light brown watery rot, which increases further. Skin becomes wrinkled and the fruit emit a peculiar musty odour.

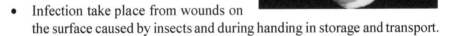

- Under humid conditions a bluish-green sporulating growth appears.

- Infection take place from wounds on the surface caused by insects and during handing in storage and transport.

Management

- Careful handling of fruits after harvest.

- Dipping the fruits aureofunginsol @ 500 ppm for 20 min.

Bitter rot (*Glomerella cingulata*)

Symptoms

- Faint, light brown discolouration beneath the skin develops. The discoloration expands in a cone shape. The circular, rough lesions become depressed. The lesions increased and cover the entire surface.

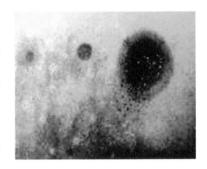

- Spray Mancozeb (0.25%) in field just before maturity.

Pear

Fire Blight (*Erwinia amylovora*)

Symptoms

- Cankers are formed on twigs and branches in the previous season.

- In the spring the bacteria begins to multiply at the same time growth starts. As the bacterium increases, ooze is formed at the margin of the canker.

- Insects are attracted to the ooze and it is carried to the open blossoms.

- Blossoms are blighted within 7-10 days after infection. After blossom infection, bacteria spread into the fruit peduncle and finally into the twig.

- Ooze is continually being produced which can add to secondary infection.

Management

- Adopt tolerant varieties like Kieffer, Orient, Garber or Douglas.

- Maintain balanced fertilizer level. Do not use excess N levels.

- Prune during dormant months. Avoid summer pruning.

- Prune 8 to 12 inches below visible sign of disease.

- Apply bactericides at five day intervals between early and late blooms.

Leaf Blight and Fruit Spot (*Entomosporium maculatum*)

Symptoms

- Spots appear as small purple marks which with age develop into purple margins with brown centres.

- Fruit spots are one-fourth inch in diameter, black, and slightly depressed. They coalesce to cover a large portion of the fruit surface.

- Secondary infection can occur during the spring and summer when the temperature is near 750F. and surface moisture is on the leaves.

Management

- Fungicides should be applied three to four times at full leaf development and continued further.

Peach

Peach leaf curl (*Taphrina deformans*)

Symptoms

- It attacks the leaves, causing curling and blister formation.
- The leaves start turning yellowish or reddish and fall off prematurely.
- The infected portion develops a pink or reddish bronze colour.
- Growth of the tree is affected with a reduction in yield.
- Pruning of and burning of infected shoots.
- Spray the plants with Bordeaux mixture 1% or 0.1% carbendazim.
- Spray Mancozeb 0.25% at 20 days before harvesting.

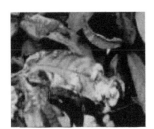

Powdery mildew (*Sphaerotheca pannosa*)

Symptoms

- Small superficial white powdery mass appear on leaves. And spread on entire plant parts.
- Fruits turn pinkish and finally dark brown in colour.

Management

- Spraying wettable sulphur (0.3%) or carbendazim (0.1%).

16

Rejuvenation of Senile Orchards

The decline of productivity and health of the tress may be due to faulty management i.e. unsuitable site and climate, cultivation of intercrops, inadequate nutrition, improper planting, undesirable planting materials, incidence of insect pest and disease and other biotic and a biotic stresses. The decline of trees are characterized by sparse appearance, yellowing and different type foliage symptoms, undergrowth and sickly appearance, dried-up top growth with small and less number of fruits. The branches of trees start to die from the top to downwards, ultimately resulted poor quality fruits. Such type of decline may be seen in whole orchards, on in a single tree or patches. Senile orchards with poor productivity are now a common problem in temperate, tropical and subtropical fruits. Since most of the fruit trees has long juvenile period, hence going for replanting, rejuvenation may be better option to convert sick tree into productive ones.

Rejuvenation Strategies

1. Mark trees and their undesired branches for pruning.

2. Pruning of marked branches should be done in the recommended month.

3. Pruning should be done in alternate row.

4. Pruning should be initiated from lower surface of the branch and alter from upper surface to avoid cracking of branch and bark splitting.

5. Immediate after heading back, apply copper oxychloride paste on the trunk, branches as well as cut surfaces to avoid infestation of diseases.

6. Care should be taken to control pests and disease.

7. After pruning more number of shoots will emerge, for proper growth of newly emerged shoots thinning out of undesired shoots is essential.

Rejuvenation of Mango Trees

In general, 40-45 years old mango trees exhibit decline in fruit yield because of dense and overcrowded canopy. The trees do not get proper sunlight resulting in decreased production of shoots. New emerging shoots are weak and are unsuitable for flowering and fruiting. The population of insects and pests built up and the incidence of diseases increases in such orchards. These unproductive trees can be converted into productive ones by pruning.

Pruning

Intermingling, diseased and dead branches are removed. Thereafter undesirable branches of unproductive trees are marked. At the end of December, these marked branches are beheaded at 1.5 to 2.0 meter from distal end and the cut portions are pasted with copper oxychloride solution.

Manuring and Fertilization

Application of 2.5 kg Urea, 3 kg Single Super Phosphate and 1.5 kg Muriate of Potash besides 120 kg well decomposed FYM is recommended. Half dose of urea with full dose of Single Super Phosphate and Muriate of Potash is applied during the end of February. The remaining half dose of urea is applied during the end of June. Full dose of FYM should be applied in the first week of July. The plants are irrigated at an interval of 15 -20 days especially during hot weather.

Unwanted emerging new shoots are regularly removed to maintain the tree canopy and avoiding overcrowding of the branches. It also helps in getting proper nourishment to retained shoots.

Mulching

Mulching at the base of pruned trees is done by using black polythene sheet (400 gauge) or heavy mulching with organic material, such as, straw, dried grass, banana leaves, immediately surrounding the main trunk drastically reduces weed growth.

Thinning of Shoots

During March-April, a number of new shoots emerge around cut portions of the pruned branches. Only 8 to 10 healthy and outward growing shoots are retained at proper distance so that a good frame-work is developed in the following years. Selective and regular thinning of shoots is essential for facilitating development of open and spreading canopy of healthy shoots. Thinning operations are undertaken during the monsoon season. Copper oxychloride fungicide (3 g / litre water) should be sprayed immediately after thinning operations.

Pest and Disease Management

Intensive care for control of insect pests and diseases is highly essential for ensuring survival of pruned trees as well as healthy growth of shoots. Infestation of stem-borer can be easily identified by wooden frass fallen on ground from the affected branches. If infestation is found, place cotton wick soaked with petrol or dichlorvos or kerosene oil.

Leaf cutting weevil damages shoot by cutting the leaf across the lamina like scissors. It can be managed by two sprays of 0.2 percent carbaryl (Sevin) insecticide (@ 3 g per litre water) at an interval of 15 days.

Brown spots on young leaves are the characteristic symptom of anthracnose disease. Copper oxychloride (3 g per litre water) should be sprayed twice at an interval of 15 days for its management.

After two years of pruning new shoots come into bearing and the yield of fruit increases gradually. Thus, old and unproductive trees are converted in to productive ones.

Rejuvenation of Citrus Trees

As spring approaches the sap starts to rise in citrus trees as their growth commences for the season. This makes it a perfect time to renovate citrus plants that may have become overgrown, Prune weak, and unproductive immediately followed by application of carbendazim spraying @ 1 gm/liter of water. Use secateurs to cut back any dead or dying shoots as well as leggy shots that have distorted or sparse leaf growth. For larger branches a pruning saw may be necessary to make clean cuts. For decline trees, cut main branches within a couple of metres of the trunk and apply a 5-10 cm deep mulch of well-rotted FYM or vermin manure and water well.

After few weeks, new shoots appearing. Thinning out shoots and keep only 8-10 healthy shoots in all direction to avoid overcrowding. Application of Dichlorovas @ 0.1% (3-5 ml) in each larval tunnel or inserting in tunnel cotton swab soaked with insecticide to control bark eating caterpillar. Scrap oozing out gum and apply Metalaxyl paste on the wound and spray Metalaxyl MZ 72 @ 2.75 gm/liter of water for the control of Phytophthora. Timely application of recommended dose of manure, fertilizer and micro-nutrients. Spraying of Imidacloprid @ 3 ml or monocrotophos @ 5 ml/ 10 liter of water for the control of citrus psylla. Application of Bordeaux paste on the tree trunk twice a year before monsoon and after monsoon. It may take a two seasons before the tree bears a really heavy crop again.

Rejuvenation of Aonla

Heading back unproductive tress and irrigate just after heading back. During the phase of heading back 50 kg FYM along with 8 kg neem cake / plant should be applied. Six months after heading back manures and fertilizers may be applied at the rate of 50 kg FYM +4 kg neem cake + 1000 g Nitrogen + 500 g potash and 750 g Phosphorus/ year. Fifty per cent of Nitrogen and entire dose of potash and phosphorus should be applied in January-February and rest dose of nitrogen is applied in June. During dry summer) regular irrigation should be given at an interval of 10-15 days.

During three to four months after pruning there is profuse mergence of shoots on pruned branches. Selective and regular thinning of shoots is essential for facilitating development of open and spreading canopy of healthy shoots. Outwardly growing 8-10 healthy shoots are retained per branch and the rest are removed so that they get proper nourishment and develop into ideal canopy. Thinning operations are undertaken during the monsoon season. Copper oxychloride fungicide (3 g / litre water) should be sprayed immediately after thinning operations. Intensive care for control of insect pests and diseases is highly essential for ensuring survival of pruned trees as well as healthy growth of shoots.

Rejuvenation of Guava

Declined guava plants should be headed back to the extent of 1.0 to 1.5 meters above the ground level during May to allow the development of fresh canopy of healthy shoots. Irrigate just after heading back. Just after heading back, 50 kg FYM along with 6 kg neem cake / plant should be applied. New shoots will emerge. These newly emerged shoots are allowed to grow up to a length of about 40 to 50 cm. These shoots were further pruned to about 50 per cent of its total length in October for emergence of multiple shoots. Six month after rejuvenation, manures 40 kg FYM + 4 kg neem cake + 1300 g urea +500 g muriate of potash and 1800 g single super phosphate / plant /year to be given. It is desired to promote fruit load in winter season. Hence, to check the onset of rainy season crop, shoot pruning (50%) was done again in May. emergence of new shoots is facilitated. These new shoots emerging after May pruning are found to have high flowering and fruiting potential for winter crop. This procedure of sequential and periodic pruning was continued every year for proper shaping of tree canopy and to ensure enhanced production of quality fruits during winter season. Adopt recommended management practices to control of pests and diseases.

Semi-declined tree of mango   Heading back   New shoots after heading back

Overcrowding of shoots after heading back   Growth of new shoots after thinning out

Rejuvenated mango orchard

17

Frost Damage: Causes and Control

Various types of winter damage can occur to fruit trees. Temperature-related injuries include sunscald, tip dieback, bud death and heartwood damage. Even after planting frost tolerant varieties and rootstocks, it is important to manage frost occurring in regions where it is suspected to happen, it is necessary to prevent injury. When temperatures dip below the freezing point, water molecules move out of plant cells through the cell membrane due to ice formation in the intercellular spaces. Other cell constituents like sugars become more concentrated within the cell and with gradual reduction in water leads to dehydration and necrosis. At relatively short and moderate freezing temperatures little damage occurs to plants, but when temperatures fluctuate drastically or descend below -4°C, ice crystals form and intracellular freezing occurs leading to cell death. Tip dieback, bud death, and heartwood damage usually occur in situations where temperatures drop dramatically before the plant has hardened off adequately.

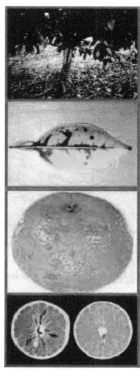

Freeze damage in Citrus

Freeze occurrence in young orchards with sprinkler and completely damaged plants.

Limiting soil moisture and the availability of nutrients late in fall, applying hormones like abscisic acid and managing air-flow may help plants dehydrate or "harden off" for winter and thereby improve survival. Sunscald occurs on the south to southwest side of trees, where the bark is exposed to sunlight during the warmest part of the day. Sunlight is absorbed by the bark and internal liquids are heated so that the cells become active; but when the sun sets and the temperature of the bark drops, the active cells die.

Types of Frost

Advection Frost

An advection frost occurs when cold air blows into an area to replace warmer air that was present before the weather change. It is associated with moderate to strong winds, no temperature inversion, and low humidity. Often temperatures will drop below 320F (00C) and stay there all day. Advection frosts are difficult to protect against as they sudden.

Radiation Frost

Radiation frosts are more common. They are characterized by clear skies, calm winds, and temperature inversions. Radiation frosts occur because of heat losses in the form of radiant energy. Under clear night skies, more heat is radiated away from an orchard than it receives, hence the temperature drops drastically. The temperature falls faster near the radiating surface causing a temperature inversion to form (temperature increases with height above the ground). This condition for certain period beyond a limit leads to damage.

Heat Transfer Methods

Movement of heat is always from a warmer to a colder section. There are three basic means of heat transfer. These include conduction, convection, and radiation.

a. **Conduction** is the transfer of heat through solid bodies in contact. Heat is transferred from molecule to molecule as it moves through the body. For example, heat movement through soil.

b. **Convection** is the transfer of heat through the movement of heated liquid or gas, such as air. For example, when a heater is operating in an orchard, cold air moves toward the heater, becomes warm, and rises upward.

c. **Radiation** is the direct transfer of heat energy through space from one object to another. This is the form by which the earth receives the sun's energy. The sun's energy traveling through space creates heat as it strikes the earth's surface. Energy radiating from the soil surface and from plants is lost to space (as on clear nights) or is absorbed or reflected by water and water vapor (as in clouds).

Differences Between a Frost and a Freeze

The terms frost and freeze are often used interchangeably but refer to two different weather events. The term freeze is normally used to describe an invasion of a large, very cold air mass from north Himalayan region. This event is also called an advective or wind-borne freeze. Wind speeds during an advective freeze are usually greater than 7 kmph. Clouds are commonly present during much or all of the event, and air is usually quite dry. Freeze protection systems are usually of limited value during this type of severe freeze. However, radiational frost (also called a radiational freeze) typically occurs when winds are calm (usually 0 to 5 kmph) and skies are clear. Under such conditions, an inversion may form because of rapid radiational cooling at the surface. If a strong inversion forms, temperatures aloft (usually up to 100 to 200 feet) may increase 7 0C degrees or more above surface temperatures.

There are two types of frosts, i.e. white frost and black frost. Visible frost occurs when atmospheric moisture freezes (forms small crystals) on plant and other surfaces. Dew (free water) forms when the air temperature equals or drops below the dew-point temperature. As temperatures continue to drop on cold nights, this dew may freeze or form frost by sunrise. Because of radiational cooling of surfaces, frost may develop on rocks, plastic, leaves, and other surfaces, while air temperature is still above freezing (0°C). If the air temperature is below the freezing point of water, when water vapour is lost from the air, ice crystals, rather than dew, form, and the frost is called white frost. The temperature at which this occurs is referred to as the frost point.

When the dew-point temperature is below the freezing temperature of the air, neither frost nor dew forms. Such a condition is referred to as a black frost. The development of frost depends on the dew point or frost point of the air. Hence, drier the air, the lower is the dew point.

Protection from Frost Damage

To prevent this problem, the bark can be painted white (or covered with a reflective wrap), so that the sun s energy is reflected away from the surfaces of the tree. Damage is more likely to occur to young plants with smooth, thin bark such as apples, pears and cherries. Protecting young trees from cold damage is a difficult task. Since, young trees are small and occupy a relatively small percentage of a planted area, protection by most active means is not particularly effective. This is especially true of heating with fossil fuel sources which are now quite expensive. Wind machines could be considered for protection, but their use is limited due to cost of acquisition and operation. Irrigation for cold protection is now widely used technique with properly designed and maintained micro-sprinkler systems. Some of the more important passive cold protection measures include cultivar and rootstock selection, site selection, clean cultivation, pre-freeze irrigation and the use of banks and wraps.

a. Passive Protection Methods

Site selection: This is the best control to avoid such crops, which are suspected to be frequently affected by frosts. At least 20 years weather data need to be studied before selecting the site. During cold nights, temperature differences are quite common in hilly areas. On radiational frost nights, as air near the surface becomes cool, denser and flows downhill to lower areas where it gets collected. These areas become much colder than those higher in the terrain. These locations are commonly called frost pockets or cold pockets. It is always better to avoid such areas. It is best to plant deciduous crops on north facing slopes to avoid cold spots at the bottom of hills and to delay springtime bloom. Probability of freezing decreases rapidly with time in the spring, and deciduous crops on south facing slopes will bloom earlier. As a result, deciduous crops on south-facing slopes are more prone to freeze damage. Subtropical trees (e.g., citrus and avocados) are damaged by freezing regardless of the season, so they are best planted on south facing slopes where the soil and crop can receive and store more direct energy from sunlight.

Wind breaks: It is advisable to plant windbreaks to avoid North and North-west freezes.

Cold tolerant varieties and rootstocks: It is the most effective strategy to mitigate the ill effects of frost. In sub-tropical and temperate region it is always better to select varieties which are proven for their tolerance to cold to avoid such a situation.

Physical Methods

Ground Cover and Mulches

When grass, weeds or any cover crop in an orchard, the sunlight is reflected from the surface as a result less energy is stored in the soil. This condition makes the crop more prone to freeze damage. Vegetative mulches usually reduce the transfer of heat into the soil and hence make crops more freeze prone. Covers are sometimes used to decrease the net radiation and convection energy losses from a crop and reduce the potential for freeze damage. The type of cover depends on the crop and the cost of labour and materials. Clear plastic mulches that increase heat transfer into the soil typically improve heat storage and hence provide passive freeze protection. Black plastic mulch is less effective for frost protection. Wetting the soil before covering with clear plastic provides the best protection.

Thatches and Polythene or Ppoly-covers

Dwarf or young trees are covered by maize or millets thatches or plastic sheets. This is done just before the sun set so that some heat can be captured. Care must be taken so that the cover material is put loosely over it. It is advisable to shield the southwest side of the tree from cool wind. Wrap should be loose and with little space for air flow.

Surface Irrigation

Surface (flood and furrow) irrigation is commonly used for freeze protection. Protection is provided by the conversion of latent to sensible heat from the cooling water. In surface irrigation, freezing of the water is undesirable because the formation of ice above the liquid water prevents heat transfer from the warmer water under the ice crust. Irrigation should be started early in the days so that the entire field is ready. Warmer water provides more protection.

Soil Banking

It is done placing a mound of soil around the tree s trunk to protect the bud union and trunk from cold. It is one of the most efficient cold protection methods for young trees. By mounding soil around the trunk of a tree (banking), heat is conducted through the soil and into the protected area of the young tree. Thus, banking protects by conduction and insulation as well. It is most efficient when the trees are banked the day before a freeze or just before the occurrence of frost is predicted. Banks can be constructed with a shovel or hoe. They are build reasonably high,
i.e. upto the scaffold limbs whenever possible. Use only soil which is devoid of weeds, and free of insect eggs.

Tree Wraps

Tree wraps are most useful in protecting young orchard trees during mild to moderate freeze. Tree wraps protect only the trunk, and consequently leaf loss can occur during moderate or severe freezes. Wraps work by delaying, but not preventing, heat loss from the tree trunk as air temperatures decrease. Wraps are most effective during freeze of short duration where temperatures drop rapidly. They are less effective, however, during freezes where temperatures decrease slowly and remain low for protracted periods. The effectiveness the wrap is related to the insulating material. When freeze damage occurs, wraps should be removed or pushed down to allow for growth of new shoots. Wraps should be properly positioned and fastened around the trunk for best results. It is important to cover the entire lower trunk, especially at the base. Selection of the proper tree wrap for a particular grove depends on a number of factors including cost, ease of installation and probability of freeze damage. For example, growers in northern regions of the state should choose wraps with good insulating qualities, while growers in warmer southern locations may opt for less costly, thinner wraps. Tree wraps also inhibit sprouts and protect trunks from herbicide and mechanical damage. Consequently, no one wrap is best for all situations.

     Foam wrap   Polyurethane wrap   Fiber-glass wrap

Foam

Advantages: It is high insulating value (3-6° above air temp), moderately durable, sprout

inhibitor, allowing for regrowth following a freeze, Inert, thus not holding water for long periods of time, rarely causing foot rot problem, moderately inexpensive, and conforms well to large or irregularly shaped trunks.

Disadvantages: More difficult to install and handle than some other wraps, Moderate ant problems.

Polyurethane Foam

Advantages: Moderate insulating value (2 - 4°C above air temp.), moderately durable, sprout inhibitor, Inert and not hold water, rarely causes foot rot problems, easy to handle and install, and Some models use irrigation water tube for extra protection.

Disadvantages: Moderately expensive, ant problems, must be removed after freeze damage to allow regrowth, may cause bark sloughing and fit poorly on large or irregularly-shaped trunks.

Microsprinkler Irrigation

Overhead, high-volume sprinklers have been used successfully in citrus nurseries for years as a means of cold protection. Recently, there has been interest in using low-volume microsprinklers to protect young trees in the field; however, success varies with the type of system, application rates, type of freeze (advective vs. radiative), and severity of the freeze.

Water protects young trees by transferring heat to the tree and the environment. The heat is provided from two sources, sensible heat and the latent heat of fusion. Most irrigation water comes out of the ground at 68° to 72°F, depending on the depth of the well. In fact, some artesian wells provide water of 80°F or more. As the water is sprayed into the air, it releases this stored (sensible) heat. However, by the time the water reaches the tree it has lost most of its energy, particularly for low volume microsprinkler systems. Consequently, the major source of heat from irrigation is provided when the water changes to ice (latent heat of fusion). As long as water is constantly changing to ice the temperature of the ice-water mixture will remain at 32°F.

The higher the rate of water application to a given area, the greater the amount of heat energy that is applied.

The major problems in the use of irrigation for cold protection occur when inadequate amounts of water are applied or under windy (advective) conditions. Evaporative cooling, which removes 7.5 times the energy added by heat of fusion, may cause severe reductions in temperature under windy conditions, particularly when inadequate amounts of water are used. In addition, most irrigation systems will not protect the upper portion of the canopy.

A number of low-volume microsprinklers which can be used for cold protection of young trees are currently available. As with tree wraps, no one system is best for a given grove situation. Remember that microsprinkler irrigation is primarily used to irrigate trees, and practical irrigation designs may not necessarily provide optimum cold protection. Again, cost, ease of operation, and especially probability of freeze damage should be considered when selecting an irrigation system. However, the key to successful cold protection using any microsprinkler system is providing a continuous and adequate volume of water directly to the trunk of the tree. This is particularly true during advective freezes where water may be blown away from the trunk.

Foggers

Natural fog is known to provide protection against freezing. Fog lines that use high pressure lines and nozzles to make fog droplets have been reported to provide excellent protection under calm wind conditions. Similarly, natural fogs created by vaporizing water with jet engines have been observed to provide protection. The jet engine approach has the advantage that it can be moved to the upwind side of the crop to be protected. However, the high pressure line approach has proven more reliable.

Heaters

The greatly increased cost of fuel has practically eliminated heaters from the grower s cold protection strategy. However, heaters can still be cost effective when used to protect high-value citrus cultivars.

Using Heaters

Orchard heaters provide heat by direct radiation and convection. Stack heaters give out 25-30 percent radiant heat, which moves along a straight line from the heater to the trees. Air around the immediate area of the heater is heated by convection; some of this heat is lost if it rises above the level of the orchard.

Because of the need for fuel-burning efficiency and pollution reduction, orchard heaters have evolved to the upright stack design. Vaporizing pot-type stack heat (for example, jumbo cones and return stacks) have the advantage of low initial cost, maneuverability, and versatility. However, fuel can be lost due to spillage, leakage, and boiling of fuel left in the heaters after they are extinguished. Labor requirements for lighting and refueling heaters are high, and an additional crew is frequently needed to refuel heaters if several nights of freeze protection are required. Compared to individual stack heaters, centralized pressure fuel systems burning diesel fuel and liquid propane are more fuel-efficient and offer considerable labor savings. Fuel storage for any heating system is a big expense and environmental liability.

Wind Machines

Wind machines offer some excellent advantages in cold protection because they minimize labor requirements, consume less fuel per acre protected and require less fuel storage than heaters. Fuel requirements for wind machines are about 12l /h. These advantages must be weighed against the disadvantages of rather high capital costs and the failure of the wind machine to provide adequate cold protection under all conditions. Wind machines are dependent on having an inversion, i.e. warmer air at approximately 40-50 feet above the orchard surface. A temperature inversion of atleast 2.5°C difference is necessary and an inversion of 3-5°C makes the wind machine very effective. They are most beneficial when located in low pockets where they mix cold, heavy air, which settles there, with warmer air above. In general, 10 horse power is required to protect one acre. Usually, one wind machine is required for each 10 acre block. However, the increase in temperatures is highest nearest the machine and decrease toward the edge of area protection. Heaters can frequently be used near the edge of the area protected to remedy this situation. Start wind machines when temperatures are two to three degrees above the lethal temperature. Because of the low cost of running a wind machine, plus the fact that it can only raise the temperature a few degrees, it is necessary to start the wind machine early. It is very important that wind machines be run effectively.

Practical

Protection of young orchards

1. Erecting thatch cover using bamboo sticks and plastic, farm trash etc.

2. Painting of trunk with Bordeaux or tree paint.

3. Cleaning of emitters giving each plant sufficient moisture during frost occurance

18

Effect of Air Pollution on Fruit Crops

Rapid urbanization and industrialization has led to continuous deterioration of air quality, which is a major environmental problems in many urban areas in both developed and developing countries. Air pollution is characterized by high concentration of suspended particulate matter, oxides of sulphur and nitrogen resulting primarily from increased use of vehicles. Air pollution effects are not limited to the short term nor to the plant damaged or killed. Rather, air pollution has long term effects that affect plants including fruit trees. Horticultural crops are injured when exposed to higher of abnormal levels of various air pollutants. The injury symptoms range from visible marks on foliage, reduced growth and yield to premature death of the plant. The development and severity of injury depends not only on the concentration of the particular pollutant, but also on a number of other supporting factors. These include the duration of exposure to the pollutant, the plant species and development stage as well as conducive environmental factors.

Air Pollution Problems

Air pollutants are injurious to vegetation can generally be classified as either local or widespread. Local pollutants are those emitted from a specific stationary source and result in a well-defined zone of vegetation injury or contamination. Most common among the local pollutants are carbon monoxide, sulfur dioxide, fluorides, ammonia and suspended particulate matter. Ozone is the most widespread pollutants, is produced in the atmosphere during a complex reaction involving nitrogen oxides and reactive hydrocarbons, components of automobile exhausts and fossil fuel combustion. Acid rain occurs as a chemical reaction fueled by the sun when air contaminants combine with the moisture. The result is rain with an acidic pH. Depending up the pH level, effects can range from plant damage to plant death, depending on concentration and period of exposure to toxins.

Effects on Plants

Air pollution injury to plants can be evident in several ways. Injury to foliage is most commonly visible in a short time and seen as necrotic lesions (dead tissue), yellowing followed by chlorosis. There is reduction in growth of various plant parts, which may ultimately lead to death of the plants due to recurrent injury. Contaminants in the air can cause the stomata of a plant/ leaf to close. Stomata is closed where toxic gas exchange occurs in the atmosphere. As result when plants cannot get adequate carbon dioxide from the air, photosynthesis ceases. Reduced growth in leaves and fruit make plants less productive. There is poor yield realization and fruits of poor appearance and quality.

Oxidants

Ozone is the major air-pollutant. Its occurrence and effect on plants are well known including fruit crops. Qzone injury to vegetation has been reported and documented in many countries across the world. Localized, domestic ozone levels contribute to the already high background levels. Injury levels vary annually and also with fruit species. Sensitive species include grape, citrus etc. Ozone tolerant species include pear, apricot etc.

Fig. 1: Ozone injury to soybean foliage

Ozone symptoms appear as appear as a flecking, bronzing or bleaching of the leaf tissues on upper leaves. Yield reductions are usually with visible foliar injury, however, crop loss can also occur without any sign of pollutant stress. Conversely, some crops can sustain visible foliar injury without any adverse effect on yield. Susceptibility to ozone injury is influenced by many environmental and plant growth factors. High relative humidity, optimum soil-nitrogen levels and water availability increase susceptibility. Injury development on broad leaves also is influenced by the stage of maturity. Young and mature leaves are resistant, while just mature leaves are most damaged.

Carbon dioxide

Carbon dioxide (CO_2) is one of the major pollutants in the atmosphere. Major sources of CO_2 are fossil fuel burning and deforestation. The concentrations of CO_2 in the air in 1860 was about 290 parts per million (ppm). In the hundred years, the concentration has increased by about 30 to 35 ppm that is by 10 percent. Industrial countries account for 65% of CO_2 emissions with the United States and Soviet Union responsible for almost 50% of this gas alone. Less developed countries (LDCs), with 80% of the world's people, are responsible for about 35% of CO_2 emissions but is expected to contribute about 50% by 2020. Carbon dioxide emissions are increasing by 4% a year.

In 1975, 18 thousand million tonnes of carbon dioxide (equivalent to 5 thousand million tons of carbon) were released into the atmosphere, but the atmosphere showed an increase of only 8 billion tonnes (equivalent to 2.2 billion tonnes of carbon. The ocean waters contain about sixty times more CO_2 than the atmosphere. When this equilibrium is disturbed by external increase in the concentration of CO_2 in the air, then the oceans would absorb more and more CO_2. If the ocean water no longer absorb, then more CO_2 remains into the atmosphere. As water warms, its ability to absorb CO_2 is further reduced. It is known that CO_2 is a good transmitter of sunlight, but partially restricts infrared radiation going back from the earth into space. This produces the so-called greenhouse effect that prevents a drastic cooling of the Earth during the night. Increasing the amount of CO_2 in the atmosphere reinforces this effect and is expected to result in a warming of the Earth's surface. Currently, carbon dioxide is responsible for 57% of the global warming trend. Nitrogen oxides contribute most of the atmospheric contaminants.

Sulfur dioxide

The major sources of sulfur dioxide in air pollution are the coal-burning operations, especially those dealing with power generation. Sulfur dioxide emissions also result from the burning of petroleum products, smelting of sulfur containing ores, vehicular population. Sulfur dioxide is a common pollutant and one of the causes of acid rain. The effects of exposure can be acute or chronic. Acute injury results in destruction of the vein network of leaves. Chronic injury, on the other hand, occurs during continuous exposure. What makes sulfur dioxide especially deadly is the fact that it can affect higher plants like trees and crops.

Sulfur dioxide enters the leaves through the stomata and the resultant injury may be either acute or chronic. Acute injury (Fig. 2) is caused by absorption of high concentrations of sulfur dioxide in a relatively short time. The symptoms appear as two-

Fig. 2: Sulfur dioxide injury to raspberry

sided (bifacial) lesions that usually occur between the veins and occasionally along the margins of the leaves. The colour of the necrotic area can vary from a light tan or near white to an orange-red or brown depending on the time of year, the plant species affected and weather conditions. Recently, expanded leaves usually are the most sensitive to acute sulfur dioxide injury, the very youngest and oldest being somewhat more resistant.

Chronic injury is caused by long-term absorption of sulfur dioxide at sub-lethal concentrations. The symptoms appear as a yellowing or chlorosis of the leaf, and occasionally as a bronzing on the under surface of the leaves. Different plant species and varieties and even individuals of the same species may vary considerably in their sensitivity to sulfur dioxide. These variations occur because of the differences in geographical location, climate, stage of growth and maturation. Crop plants which are considered susceptible to sulfur dioxide are alfalfa, barley, buckwheat, clover, oats, pumpkin, radish, rhubarb, spinach, squash, Swiss chard and tobacco. Resistant crop plants include asparagus, cabbage, celery, corn, onion and potato.

Fluorides

Fluorides are discharged in air due to combustion of coal; the production of brick, tile, enamel frit, ceramics and glass; the manufacture of aluminium and steel; and the production of hydrofluoric acid, phosphate chemicals and fertilizers etc. Fluorides absorbed by leaves are conducted towards the margins of broad leaves (grapes). Little injury takes place at the site of absorption, whereas the margins or the tips of the leaves build up injurious concentrations. The injury (Fig. 3) starts as a gray or light-green water-soaked lesion, which turns tan to reddish-brown. With continued exposure the necrotic areas increase in size, spreading inward to the mid-rib on broad leaves and downward on monocotyledonous leaves.

The fluoride enters the leaf through the stomata and is moved to the margins, where it accumulates and causes tissue injury. Note, the characteristic dark band separating the healthy (green) and injured (brown) tissues of affected leaves. Susceptibility of plant species to fluorides show that apricot, blueberry, peach, gladiolus, grape, plum, prune, etc. are most sensitive. Tolerant types include pear, strawberry etc.

Fig. 3: Fluoride injury on plum foliage

Ammonia

Ammonia injury is common in areas in vicinity of fertilizer industries manufacturing anhydrous and aqua ammonia fertilizers. The units do release large quantities of ammonia into the atmosphere, which cause severe injury to plants growing in vicinity. Repeated exposure leads to appearance of irregular, bleached, bifacial, necrotic

Fig. 4: Severe ammonia injury to apple foliage

lesions. In the case of severe injury to fruit trees, recovery through the production of new leaves can occur (Fig.4). Sensitive species include apple, raspberry etc. Tolerant species include alfalfa, beet, carrot, corn, cucumber, eggplant, onion, peach, rhubarb and tomato.

Chlorofluorocarbons (CFCs)

CFCs are lowering the average concentration of ozone in the stratosphere. Since 1978, the use of CFCs in aerosol has been banned in countries like USA, Canada, and several Scandinavian countries. Aerosols are still the largest use, accounting for 25% of global CFC use. Spray cans, discarded or leaking refrigeration and air conditioning equipment, and the burning plastic foam products release the CFCs into the atmosphere. Depending on the type, CFCs stay in the atmosphere from 22 to 111 years. Chlorofluorocarbons move up to the stratosphere gradually over several decades. Under high energy ultra violet (UV) radiation, they breakdown and release chlorine atoms, which speed up the breakdown of ozone ($O3$) into oxygen gas ($O2$).

Chlorofluorocarbons, also known as Freons, are greenhouse gases that contribute to global warming.

Suspended particulate matter

Particulate matter such as cement dust, magnesium-lime dust and carbon soot deposited on vegetation can inhibit the normal respiration and photosynthesis mechanisms within the leaf. Cement dust may cause chlorosis and death of leaf tissue by the combination of a thick crust and alkaline toxicity produced in wet weather. The dust coating (Fig. 5) also may affect the normal action of pesticides and other agricultural chemicals applied as sprays to foliage. In addition, accumulation of alkaline dusts in the soil can increase soil pH to levels adverse to crop growth.

Fig. 5: Cement-dust coating on apple leaves and fruit.

The dust had no injurious effect on the foliage, but inhibited the action of a pre-harvest crop spray which washes it.

Smog

Photochemical air pollution is commonly referred to as "smog". Smog, a contraction of the words smoke and fog, has been caused throughout recorded history by water condensing on smoke particles, usually from burning coal. With the introduction of petroleum to replace coal economies in countries, photochemical smog has become predominant in many cities, which are located in sunny, warm, and dry climates with many motor vehicles. Photochemical smog is also appearing in regions of the tropics and subtropics where grasses and weeds are periodically burned. Smog's unpleasant properties result from the irradiation by sunlight of hydrocarbons caused primarily by unburned gasoline emitted by automobiles and other combustion sources. The products of photochemical reactions include organic particles, ozone, aldehydes, ketones, peroxyacetyl nitrate, organic acids, and other oxidants. Ozone is a gas created by nitrogen dioxide or nitric oxide when exposed to sunlight. Ozone causes eye irritation, impaired lung function, and damage to trees and crops.

Another form of smog is called Industrial Smog. This smog is created by burning coal and heavy oil that contain sulfur impurities in power plants, industrial plants, etc. The smog consists mostly of a mixture of sulfur dioxide and fog. Suspended droplets of sulfuric acid are formed from some of the sulfur dioxide, and a variety of suspended solid particles. When power plants burned a large amount of coal and heavy oil leading to large-scale problems. The highly dense of smog caused human and plant death.

Air pollution has caused reduction in production and quality deterioration in different fruit crops. It has been evident in regions where big industrial establishments exist or coming up in the fruit growing regions due to rapid industrialization.

Precautions to be adopted by Fruit Growers

1. Avoid areas where air polluting units like industries exist.

2. Periodically spray water to clean the leaves of deposits and dirt.

3. Plant high wind break to avoid movement of smog.

4. Remedial measures must be applied immediately to avoid death of plants.

Colour Plate Section of Part-II: Practicals

1: Identification of Fruit Varieties

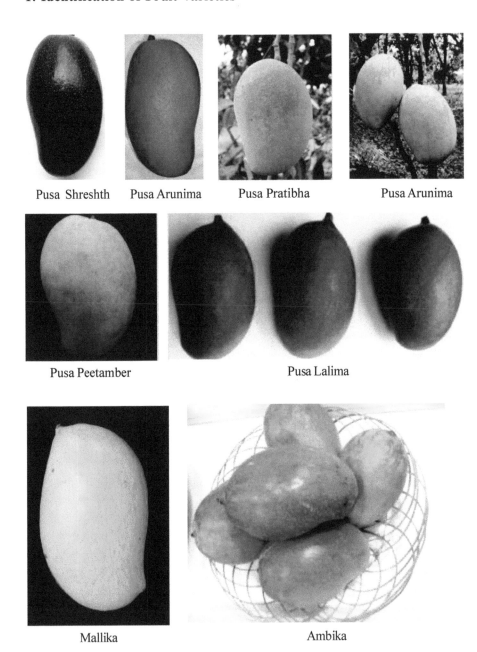

Pusa Shreshth   Pusa Arunima   Pusa Pratibha   Pusa Arunima

Pusa Peetamber   Pusa Lalima

Mallika   Ambika

Malta

Mosambi

Kagzi lime

Kagzi Kalan

Galgal

Sweet lemon

Red Blush

Marsh Seedless

Calamondin   Attani

Carrizo citrange   Attani

Jatti Khatti   Rangpur lime   Trifoliate orange

Pusa Nanha   Pusa Dwarf   Red Lady

T-1   CO-5

Thar Sevika   Thar Bhubhraj

15: Major Diseases of Fruit Crops and Their Remedial Measures

17: Frost Damage: Causes and Control

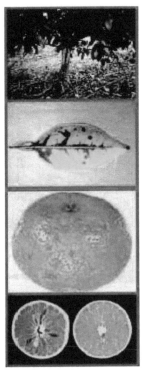

Freeze occurrence in young orchards with sprinkler and completely damaged plants.

Freeze damage in Citrus

Soil banking

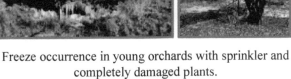

Foam wrap   Polyurethane wrap   Fiber-glass wrap

18: Effect of Air Pollution on Fruit Crops

Fig. 1: Ozone injury to soybean foliage

Fig. 2: Sulfur dioxide injury to raspberry

Fig. 3: Fluoride injury on plum foliage

Fig. 4: Severe ammonia injury to apple foliage

Fig. 5: Cement-dust coating on apple leaves and fruit.

Lightning Source UK Ltd.
Milton Keynes UK
UKHW021851041220
374638UK00001B/27